D. Schmähl C. Thomas R. Auer

Iatrogenic Carcinogenesis

With 27 Tables

Springer-Verlag Berlin Heidelberg New York 1977

Professor Dr. Dietrich Schmähl
Deutsches Krebsforschungszentrum
Institut für Toxikologie und Chemotherapie
Im Neuenheimer Feld 280, D-6900 Heidelberg

Professor Dr. Carlos Thomas
Pathologisches Institut der Universität Freiburg
Albertstr. 19, D-7800 Freiburg i. Br.

Dr. Rolf Auer
Pathologisches Institut der Universität Freiburg
Albertstr. 19, D-7800 Freiburg i. Br.
Federal Republic of Germany

ISBN-13:978-3-540-08200-2   e-ISBN-13:978-3-642-66630-8
DOI: 10.1007/978-3-642-66630-8

Library of Congress Cataloging in Publication Data. Schmähl, Dietrich, 1925 – Iatrogenic carcinogenesis. Includes bibliographical references. 1. Carcinogenesis. 2. Iatrogenic diseases. I. Thomas, Carlos, joint author. II. Auer, Rolf, 1947 – joint author. III. Title. RC 268.5. S 35 616.9'94'071   77-3933

2121/3321-543210

Contents

Preface

The purpose of this book, which is written mainly for pharmacotherapists, is to draw the physician's attentions, through specific examples, to iatrogenic (i.e., diagnostically or therapeutically induced) carcinogenesis. The book is not intended, however, to arouse public concern.

It has been known for some time that each therapeutic measure has a risk-benefit ratio, the estimation of which requires an understanding of carcinogenic side-effects. The number of iatrogenic tumors published so far is not large; however, we assume that there are a number of cases that have not been registered. In the past it was not generally recognized that medical treatment might involve carcinogenic risks; moreover, various therapeutic measures are often used simultaneously so that it may be difficult to trace the cause of tumor formation to a specific agent.

Animal experiments and clinical observations during the past few years have contributed to our knowledge of the possible hazards of drugs. We have compiled the most important literature on the subject without, however, attempting completion. The present book may help the pharmacotherapist balance the benefit of a drug against its potential risk, and is thus consistent with the medical principle of *nil nocere*.

We thank Mrs. M. Gomille (Institut für Toxikologie und Chemotherapie, Deutsches Krebsforschungszentrum, Heidelberg) for her help in the preparation of the manuscript.

Heidelberg, August 1976   D. Schmähl, C. Thomas, R. Auer

Introduction

In recent years iatrogenic carcinogenesis has become an important field of iatrogenic pathology (Geyer, 1955; Druckrey, 1956; Schmähl, 1959; 1972; Uehlinger, 1961; Olbert, 1962; Lampert, 1964; Spain et al., 1967; Truhaut, 1967; Belpomme et al., 1972; Thomas, 1972; Goerttler, 1972; Karrer and Wrba, 1972; Clayson, 1972; Waldenström, 1974; Deichmann, 1975; Preussmann, 1975). It is difficult to establish a relation between a specific therapy and the occurrence of a tumor, because these tumors often have latency periods of several years or even decades and because man is exposed continuously to various environmental carcinogenic factors. Therefore a syncarcinogenic action of both a specific drug and various other carcinogens has to be considered in most cases.

Our knowledge of the etiology and the pathogenesis of most human tumors is still not complete. It has been possible only in a few types of human tumors to determine that a specific carcinogen is responsible for the effect. Experimental results, although they cannot be extrapolated unconditionally to human conditions, may contribute to our knowledge of the origin of human tumors. It is important to publish observations of iatrogenic cancers, especially since in these cases it is possible to trace back the dose, duration of treatment, and mode of application of a drug. From this point of view an iatrogenic tumor is an unwanted experiment in man.

We have compiled about 1000 cases of iatrogenic tumors from the fields of internal medicine and surgery and their marginal areas. Tumors associated with diagnostic and therapeutic radiation will not be considered here since comprehensive reviews on this subject have already been published (Totten et al., 1957; Phillips and Sheline, 1963; Pack and Davis, 1965; Wenz, 1964; Silva Horta et al., 1965; Wegener et al., 1971; Modan and Lilienfeld, 1964; Sheline et al., 1959; Staffurth, 1966; Burke et al., 1967; Baker, 1969; Araki and Oshiro, 1970; McDougall et al., 1971; Miller, 1972; Mole, 1973; Bross and Natarajan, 1974; Seydel, 1975).

The following questions will be considered in this book:
1. Which drugs, according to the scientific literature, have to be considered as potential carcinogens in man?

2. Which criteria are valid in determining potential carcinogenic risks by medical treatment?
3. Which drugs, showing carcinogenic effects in experimental animals, have to be considered as potential carcinogens in man as well?

References

Araki, M., Oshiro, K.: Papillary adenocarcinoma of the thyroid developing after treatment of hyperthyroidism with 131 I. Gann *61*, 267 (1970)

Baker, H. W.: Anaplastic thyroid cancer twelve years after radioiodine therapy. Cancer (Philad.) *23*, 885 (1969)

Belpomme, D., Blondel, P., Mathé, G.: Les cancers d'origine thérapeutique. Vie. Med. Canada Franc. *1*, 140 (1972)

Bross, D. J., Natarajan, N.: Risk of leukemia in suspectible children exposed to preconception, in utero and postnatal radiation. Prev. Med. *3*, 361 (1974)

Burke, G., Levinson, M. J., Zitman, I.: Thyroid carcinoma ten years after sodium iodide I 131 treatment. J. Amer. Med. Ass. *199*, 247 (1967)

Clayson, D. B.: Carcinogenic hazards due to drugs. In: Drug-Induced Diseases. Meyler, L., Peck, H. M. (eds.) Amsterdam: Excerpta Medica, 1972, Vol. IV, pp. 91–109

Deichmann, W. B.: Cummings Memorial Lecture, 1975. J. Amer. Ind. Hyg. Ass. *33*, 411 (1975)

Druckrey, H.: Krebserzeugende Eigenschaften bei Arzneimitteln. Muench. med. Wschr. *98*, 295 (1956)

Geyer, A.: Zur Frage der iatrogenen Krebse. Z. inn. Med. *10*, 685 (1955)

Goerttler, K.: Modell und Wirklichkeit der iatrogenen Carcinogenese. Verh. Dtsch. Ges. Path. *56*, 138 (1972)

Karrer, K., Wrba, H.: Bedeutung der iatrogenen Cancerogenese. Verh. Dtsch. Ges. Path. *56*, 164 (1972)

Lampert, F.: Iatrogene Schäden. Med. Klin. *59*, 1001 (1964)

McDougall, I. R., Kennedy, J. S., Thomson, J. A.: Thyroid carcinoma following iodine-131 therapy. Report of a case and review of the literature. J. Clin. Endocr. *33*, 287 (1971)

Miller, R. W.: Radiation induced cancer. J. Nat. Cancer Inst. *49*, 1221 (1972)

Modan, B., Lilienfeld, A. M.: Leukaemogenic effect of ionising-irradiation treatment in polycythaemia. Lancet *2*, 439 (1964)

Mole, R. H.: Late effects of radiation: Carcinogenesis. Brit. Med. Bull. *29*, 78 (1973)

Olbert, Th.: Iatrogene Karzinogene. Krebsarzt *17*, 154 (1962)

Pack, C. T., Davis, J.: Radiation cancer of the skin. Radiology *84*, 436 (1965)

Phillips, T. L., Sheline, G. E.: Bone sarcoma following radiation therapy. Radiology *81*, 992 (1963)

Preussmann, R.: Chemische Carcinogene in der menschlichen Umwelt. In: Handbuch der allgemeinen Pathologie. Geschwülste/Tumors II. Altmann, H.-W., Büchner, F., Cottier, H., Grundmann, E., Holle, G., Letterer, E., Masshoff, W., Meesen H., Roulet, F., Seifert, G., Siebert, G. (eds.) Berlin-Heidelberg-New York: Springer, 1975, Vol. 6, Part 6, pp. 421–594

Schmähl, D.: Gibt es krebserzeugende Arzneimittel? Mentor Medici *18*, 343 (1959)

Schmähl, D.: Toxikologische Probleme der iatrogenen Carcinogenese. Verh. Dtsch. Ges. Path. *56*, 133 (1972)

Seydel, H. G.: The risk of tumor induction in man following medical irradiation for malignant neoplasm. Cancer (Philad.) *35*, 1641 (1975)

Sheline, G. E., Lindsay, S., Bell, H. G.: Occurrence of thyroid nodules in children following therapy with radioiodine for hyperthyroidism. J. Clin. Endocr. *19*, 127 (1959)

Silva Horta, D. A. J., Abatt, J. D., Cayolla da Motta, L., Roriz, M. L.: Malignancy and other late effects following administration of thorotrast. Lancet *2*, 201 (1965)

Spain, D. M., Kowal, S. J., Lunger, M., Shapiro, A., Slatkin, M. R., Warwhaw, L. J.: Iatrogene Krankheiten. Stuttgart: Georg Thieme 1967

Staffurth, J. S.: Thyroid cancer after 131 I therapy for thyrotoxicosis. Brit. J. Radiol. *39*, 471 (1966)

Thomas, C.: Systematik der iatrogenen Cancerogenese. Verh. Dtsch. Ges. Path. *56*, 126 (1972)

Totten, R. S., Antypas, P. G., Supertius, S. M., Gaisford, J. C., White, W. L.: Pre-existing roentgen-ray dermatitis in patients with skin cancer. Cancer (Philad.) *10*, 1024 (1957)

Truhaut, R. (ed.): Potential carcinogenic hazards from drugs. UICC Monograph Series, Vol. VII, Berlin-Heidelberg-New York: Springer 1967

Uehlinger, E.: Pathologische Anatomie der Therapieschäden. Verh. Dtsch. Ges. inn. Med. *67*, 457 (1961)

Waldenström, J. G.: Iatrogene maligne Zustände. Med. Welt *25*, 669 (1974)

Wegener, K., Wesch, H., Kampmann, H.: Retikuloendotheliales System und Thorotrastose nach diagnostischer Angiographie. Dtsch. med. Wschr. *96*, 1977 (1971)

Wenz, W.: Thorotrasttumoren. Quantitative Untersuchungen über das Dosis-Wirkungsproblem bei Thorotrastose. Ergebn. Chir. *46*, 81 (1964)

Examples of Iatrogenic Carcinogenesis in the Field of Internal Medicine and its Marginal Areas

Arsenic

The oncogenic effects of inorganic arsenic compounds in man are well known. Paris (1820) described occupational skin cancers following exposure to arsenic, and v. Pein (1943), Liebegott (1949), Roth (1956), Hess (1956), Galy et al. (1963), Thiers et al. (1967), and Denk et al. (1969) described tumors of the skin, bronchi, and liver in German and French winegrowers. Carcinomas caused by the arsenic content of the drinking water were observed in Reichenstein, Silesia (Geyer, 1898), Cordoba, Argentinia (Arguello et al., 1939), and on the southwest coast of Taiwan (Yeh et al., 1968)[1]. Experimental studies on arsenic carcinogenicity revealed contradictory results (Askenazy, 1926; Büngeler, 1958; Knoth, 1966; Osswald and Goerttler, 1971).

Hutchinson (1887), Ullmann (1930), and Neubauer (1947) described tumors following medication with inorganic arsenic. The salts and the anhydride of the arsenous acid in particular were given per os and less frequently parenterally for the treatment of dermatoses, anemia, epilepsy, syphilis, bronchial asthma, and general debilities. The arsenous acid blocks the sulfhydryl enzyme system and thus inhibits cellular metabolism (Goodman and Gilman, 1970). Chronic arsenic poisoning includes palmoplantar hyperkeratoses, melanoses, and inflammation of the mucosa of the nose, eye, larynx, and gastrointestinal tract.

Table 1 contains a compilation of 226 cases with arsenic-related carcinoma, including 135 men and 62 women; in 28 cases the sex was not given. The age of the patients ranged from 22 to 75 years (average age: 50 years). Basic diseases were described in 196 cases; these are shown in Table 2. Arsenic was used predominantly for the treatment of psoriasis vulgaris. The doses applied, which were seldom stated exactly, varied considerably. Total doses ranged from 0.19 g to several hundred grams. In 66 cases the duration of treatment was given exactly; in these the mean latency period of the tumors was 21 years. In the remaining cases only general data were given (i. e., "treatment for a long period of time," "treatment for several years," "treatment continued since the childhood"). In 167 cases typic arsenical keratoses of the palmo-

---

[1] For further reviews on arsenic carcinogenesis see Finsterer (1922), Fischer (1939), Traub (1939), Kleine-Natrop (1962), Bauer (1963)

plantar region were observed, which occurred mainly in those parts of the body not exposed to sunlight. In 12 cases arsenical keratoses were excluded, and in 47 cases no statements were given. Histologic diagnoses were given in 143 cases (Table 4). Squamous cell and basal cell carcinomas appeared most frequently. Nineteen of these patients had visceral tumors also (Table 5). Eleven patients, five of them with arsenical keratoses, had visceral tumors exclusively (Table 6).

The age of the patients was below that observed for other skin carcinomas (Neumann and Schwank, 1960; Fierz, 1966; Denk et al., 1969). The sex ratio of 2:1 in favor of males accords with general descriptions (Neubauer, 1947). Several authors abserved multiple skin tumors following medication with inorganic arsenic (Ullmann, 1963; Gottron, 1954; Neumann and Schwank, 1960; Fierz, 1966). In the present study multiple tumors were formed in 165 patients (63%). The incidence of multiple tumors is usually only 4–6% (Hueper, 1942; Neumann and Schwank, 1960).

The skin carcinomas occurred predominantly on the trunk, palms, feet, and extremities, whereas the common site of skin tumors is the head (Miescher, 1963; Yeh et al., 1968). The same incidences were observed for basal cell and squamous cell carcinomas. Fierz (1966) found a distinct predominance of basal cell carcinomas among 21 patients treated with arsenic.

The risk of cancer following treatment with inorganic arsenic is illustrated by the observations of Fierz (1966) who found that 40% of 262 patients treated with Fowlers solution developed palmoplantar hyperkeratoses, and 8% had skin carcinomas. He demonstrated that the cancer rate increases with increasing doses, and that there is no "harmless" dose. In several cases he observed tumors following the application of 1 g arsenic or less.

The pathogenesis of the arsenic-related carcinomas is so far unclear. According to Warburg's theory, which is justifiably disputed today, chronic arsenic poisoning irreversibly damages respiration, thereby increasing fermentation metabolism (Warburg, 1956). This hypothesis was confirmed experimentally by Büngeler (1958), who found a clear inhibition of respiration and an increase in glucolase in mice with chronic arsenic poisoning. Jung and Trachsel (1970), in a study of living epidermic cells, found that both the cytostatic and the carcinogenic effects of inorganic arsenic compounds were caused by a primary change in the DNA molecule. These investigators thought that arsenic inhibited different enzymatic systems and in particular DNA polymerization. Consequently, DNA replication and mitotic activity are inhibited; the repair mechanisms of the cell nuclei are interrupted, and the cell is thus exposed unprotected to various carcinogenic factors (i.e., UV light, chemical carcinogens, viruses). Long-standing lesions increase the risk of the early formation of multiple tumors. Since all cellular systems may be afflicted, arsenical tumors may be formed in different organs.

Although many clinicians take the view that there is no longer a need to use arsenic in dermatology and internal medicine, further increase in the incidence of arsenical cancer is expected since, according to Ehlers (1968), 73% of the practicing German dermatologists prescribe arsenicals.

References

Alderson, H. E. (1935): cited by Neubauer, O. (1947)

Alexander, A.: Arch. Derm. Syph. (Chicago) *129*, 5 (1921)

Aliferis (1924): cited by Ullmann, K. (1930)

Amersbach, J.: Arch. Derm. (Wien) *65*, 505 (1952)

Anderson, N. P.: Bowen's precancerous dermatosis and multiple benign superficial epithelioma. Evidence of arsenic as an etiologic agent. Arch. Derm. Syph. (Chicago) *26*, 1052 (1932)

Anderson, N. P. (1937): cited by Neubauer, O. (1947)

Anderson, N. P.: cited by Goecermann, W. H., Wilhelm, L. F. X. (1940)

Anderson, N. P. (1943a): cited by Neubauer, O. (1947)

Anderson, N. P. (1943b): cited by Neubauer, O. (1947)

Andrews, N. P. (1932): cited by Neubauer, O. (1947)

Applestein, R. (1941): cited by Neubauer, O. (1947)

Arguello, R. A., Cenget, D. D., Tello, E. E.: Cancer and regional edemic chronic arsenicism. Brit. J. Derm. *51*, 548 (1939)

Arhelger, S. W., Kremen, A. J.: Arsenical epitheliomas of medicinal origin. Surgery *30*, 977 (1951)

Askanazy, M.: Über den Einfluß des Arsens auf verpflanztes embryonales Gewebe. Verh. dtsch. Ges. Path. *21*, 182 (1926)

Ayres, S.: Chronic ulceration of the soles: multiple benign superficial epitheliomas; arsenical keratoses (palms, soles, trunk). Arch. Derm. Syph. (Chicago) *29*, 613 (1934)

Ayres, S. (1935): cited by eubauer, O. (1947)

Ayres, S. (1937): cited by Anderson, N. P. (1937)

Ayres, S. (1943): cited by Neubauer, O. (1947)

Barber, H. W. (1924): cited by Ullmann, K. (1930)

Barber, H. W.: Brit. J. Derm. *45*, 475 (1933)

Barber, H. W.: Brit. J. Derm. *58*, 85 (1939)

Barney, L. (1944): cited by Neubauer, O. (1947)

Bartak, P., Kejda, J.: Arsenbasaliom der Haut in jugendlichem Alter. Hautarzt *23*, 457 (1972)

Bauer, K. H.: Das Krebsproblem, 2nd ed. Berlin-Heidelberg-New York: Springer 1963

Berlin, C., Tager, A.: Psoriasis complicated by cutaneous and visceral carcinomatosis due to ingestion of arsenic. Acta derm. venereol. (Stockh.) *42*, 252 (1962)

Bland-Sutton, J. (1916): Brit. Med. J. *2*, 788 (1916)

Bloom, D. (1936): cited by Neubauer, O. (1947)

Brocq (1902): cited by Neubauer, O. (1947)

Braun, W.: Krebs an Haut und inneren Organen, hervorgerufen durch Arsen. Dtsch. med. Wschr. *83*, 870 (1958)

Bruusgard (1928): cited by Ullmann, K. (1930)

Büngeler, W.: Der Arsenkrebs. Muench. med. Wschr. *100*, 1117 (1958)

Cartaz, A. (1877): cited by Ullmann, K. (1930)

Cole and Driver (1929): cited by Ullmann, K. (1930)

Crocker, H. R., Pernet, G. (1901): cited by Ullmann, K. (1930)

Darier, J. (1902): cited by Ullmann, K. (1930)

Degos, R., Delzant, O., Hewitt, J. (1950): cited by Unna, P. J., Memmersheimer, A., Herzberg, J.J. (1963)

Denk, R., Holzmann, H., Lange, H.-J., Greve, D.: Über Arsenspätschäden bei obduzierten Moselwinzern. Med. Welt *20*, 557 (1969)

Doepfmer, R. (1959/60): cited by Unna, P. J., Memmersheimer, A., Herzberg, J. J. (1963)

Doty, C. A. (1931): cited by Neubauer, O. (1947)

Dowling, G. B.: Proc. roy. Soc. Med. *38*, Sect. Derm. 127 (1945)

Dubreuilh, W. (1910): cited by Ullmann, K. (1930)

Ebert, M. H., Otsuka, M. (1943): cited by Neubauer, O. (1947)

Ehlers, G.: Klinische und histologische Untersuchungen zur Frage arzneimittelbedingter Arsen-Tumoren. Z. Hautkr. *43*, 763 (1968)

Fassrainer, S. (1937): cited by Neubauer, O. (1947)

Fierz, U.: Katamnestische Untersuchungen über die Nebenwirkungen der Therapie von Hautkrankheiten mit anorganischem Arsen. Arch. klin. exp. Derm. *227*, 286 (1966)

Finsen (1927): cited by Ullmann, K. (1930)

Finsterer, H.: Med. Klin. *1*, 549 (1922)

Fischer, W. (1932): cited by Neubauer, O. (1947)

Fordyce and McKee (1914): cited by Ullmann, K. (1930)

Franseen, C. C., Taylor, G. W.: Arsenical keratoses and carcinomas. Amer. J. Cancer *22*, 287 (1934)

Fuhs (1929): cited by Ullmann, K. (1930)

Galy, P., Touraine, R., Brune, J., Rouchier, P., Gallois, P.: Le cancer poulmonaire d'origin arsenicale des vignerons de Beaujolais. J. Franc. Méd. Chir. thor. *17*, 303 (1963)

Gelbjerg-Hansen (1924): cited by Ullmann, K. (1930)

Geyer, L.: Über die chronische Hautveränderung beim Arsenicismus und Betrachtungen über die Massenerkrankungen in Reichenstein in Schlesien. Arch. Derm. Syph. (Berl.) *43*, 221 (1898)

Goeckermann, W. H., Wilhelm, L. F. X.: Arsenic as cause of cancer of mucous membrane; report of case. Arch. Derm. Syph. (Chicago) *42*, 641 (1940)

Goodman, L. S., Gilman, A.: The pharmacological basis of therapeutics. New York: The Macmillan Co., 1970

Gottron, H. A.: Gegenwartsfragen beim Hautkarzinom. Med. Klin. *49*, 1553 (1954)

Gray, A. M. E. (1912): Brit. J. Derm. *24*, 325 (1912)

Gross, P. (1931): cited by Neubauer, O. (1947)

Guggenheim, L.: Multiple Präcancerosen (mit präcancerösem Exanthem) und Carcinome, zum großen Teil von bowenoiden Typus nach langjährigem Arsengebrauch. Arch. Derm. Syph. (Berl.) *168*, 26 (1933)

Haagensen, C. D.: Amer. J. Cancer *15*, 656 (1931)

Haldin-Davis (1939): cited by Neubauer, O. (1947)

Hall, A. F.: Arch. Derm. *50*, 148 (1944)

Hamilton, S. R. (1921): cited by Ullmann, K. (1930)

Harbitz, H. F. (1927): cited by Neubauer, O. (1947)

Hartzell, M. B. (1906): cited by Ullmann, K. (1930)

Hartzell, M. B., Stelwagon, H. W. (1899): cited by Ullmann, K. (1930)

Haxthausen, H.: Hospitalstidende *44*, 1 (1923)

Haxthausen, H. (1928): cited by Ullmann, K. (1930)

Hebra, H. (1897): cited by Ullmann, K. (1930)

Herxheimer (1926): cited by Ullmann, K. (1930)

Hess (1956): cited by Bauer, K. H. (1963)

Heuss (1902): cited by Ullmann, K. (1930)

Hofmann, E. (1928): cited by Ullmann, K. (1930)

Hohmann, W. J.: Ned. T. Geneesk. *86*, 1408 (1942)

Hövelborn, K.: Karzinomentstehung auf chronischen Dermatosen (Psoriasis, Ulcus cruris und Lichen ruber verrucosus, spätexsudatives Ekzematoid). Derm. Wschr. *101*, 858 (1935)

Hueper, W. C.: Morphological aspects of experimental actinic and arsenic carcinomas in skin of rats. Cancer Res. *2*, 551 (1942)

Hutchinson, J.: Arsenic cancer. Brit. med. J. *2*, 1280 (1887)

Hutchinson, J. (1888, 1894, 1898, 1902, 1903): cited by Ullmann, K. (1930)

Hutchinson, J., Albutt, C. (1888): cited by Ullmann, K. (1930)

Hutchinson, J., Bullock, J. E. (1898): cited by Ullmann, K. (1930)

Hutchinson, J., Tay, W. (e898): cited by Ullmann, K. (1930)

Hyde, J. N. (1899): cited by Ullmann, K. (1930)

Jung, E. G., Trachsel, B.: Molekularbiologische Untersuchungen zur Arsencarcinogenese. Arch. klin. exp. Derm. *237*, 819 (1970)

Kleine-Natrop, H. E.: Psoriasiscarcinom. Hautarzt *10*, 224 (1959)

Kleine-Natrop, H. E.: Multiple Malignome und Hyperkeratosen bei arsenbehandelter Psoriasis. Derm. Wschr. *145*, 390 (1962)

Knoth, W.: Arsenbehandlung. Arch. klin. exp. Derm. *227*, 228 (1966)

Konrad (1928): cited by Ullmann, K. (1930)

Lane, W. A. (1894): cited by Ullmann, K. (1930)

Laymon, C. W. (1943): cited by Neubauer, O. (1947)

Levin, O. L.: Arch. Derm. Syph. (Chicago) *13*, 569 (1926)

Liebegott, G.: Pathologische Anatomie der chronischen Arsenvergiftungen. Dtsch. med. Wschr. *74*, 855 (1949)

Löwenberg, M. (1913): cited by Ullmann, K. (1930)

Madsen, A. (1941): cited by Neubauer, O. (1947)

Mayer, L. (1928): cited by Ullmann, K. (1930)

McCormac, H.: Arsenical keratoses: multiple basal cell carcinomata and arsenical "warts." Proc. roy. Soc. Med. *26*, 1553 (1933)

McNeer, G.: Arsenical keratoses and epitheliomas. Ann. Surg. *99*, 348 (1934)

Meyhöfer, W., Knoth, W.: Über die Auswirkungen einer langjährigen antipsoriatrischen Arsentherapie auf mehrere Organe unter besonderer Berücksichtigung andrologischer Befunde. Hautarzt *17*, 309 (1966)

Miescher (1963): cited by Fierz, U. (1966)

Milian, G. (1931): cited by Neubauer, O. (1947)

Minkowitz, S.: Multiple carcinomata following ingestion of medicinal arsenic. Ann. Intern. Med. *61*, 296 (1964)

Montgomery, H.: Arsenic as an etiologic agent in certain types of epithelioma. Arch. Derm. *32*, 218 (1935)

Montgomery, H., Waisman, M.: Epithelioma attributable to arsenic. J. invest. Derm. *4*, 365 (1941)

Montgomery, R. M. (1942): cited by Neubauer, O. (1947)

Nander (1923): cited by Ullmann, K. (1930)

Neubauer, O.: Arsenical cancer: A review: Brit. J. Cancer *1*, 192 (1947)

Neumann, E., Schwank, R.: Multiple malignant and benign epidermal and dermal tumours following arsenic. Acta. derm. venereol. (Stockh.) *40*, 400 (1960)

Nicolas, Gate and Lebeuf (1923): cited by Ullmann, K. (1930)

Novey, H. S., Martel, S. H.: Asthma, arsenic, and cancer. J. Allergy clin. Immunol. *44*, 315 (1969)

Oliver (1923): cited by Ullmann, K. (1930)

Oliver and Finnerud (1928): cited by Neubauer, O. (1947)

Osswald, H., Goerttler, K.: Leukosen bei der Maus nach diaplazentarer und postnataler Arsen-Applikation. Verh. dtsch. Ges. Path. *55*, 289 (1971)

Paris, J. A.: Pharmacologica. London: Philips, 1820, 3 rd ed.

Parkhurst, H. J.: cited by Waugh, J. F., Scull, C. W. (1935)

Peck, S. M. (1942): cited by Neubauer, O. (1947)

v. Pein, H.: Über die Krebsentstehung bei der chronischen Arsenvergiftung. Dtsch. Arch. klin. Med. *190*, 429 (1943)

Piper, H. G.: Besondere Verlaufsformen der superfiziellen Epitheliomatose. Derm. Wschr. *112*, 429 (1941)

Pozzi, S. (1874): cited by Ullmann, K. (1930)

Pusey (1919): cited by Neubauer, O. (1947)

Pye-Smith, R. J., Hardwoodnutt, W. (1913): cited by Ullmann, K. (1930)

Ramel (1929): cited by Ullmann, K. (1930)

Rasch, C. (1930): cited by Neubauer, O. (1947)

Rasch and Fons (1920): cited by Neubauer, O. (1947)

Rauschkolb (1938): cited by Neubauer, O. (1947)

Regelson, W., Kim, U., Ospina, J., Holland, J. F.: Haemangioendothelial sarcoma of liver from chronic arsenic intoxication by Fowler's solution. Cancer (Philad.) *21*, 514 (1968)

Robba, G. (1920): cited by Unna, P. J., Memmersheimer, A., Herzberg, J. J. (1963)

Robinson, S. (1935): cited by Neubauer, O. (1947)

Robinson, S.: Arch. Derm. *36*, 870 (1937)

Robinson, S. (1941): cited by Rodgers, J. D. (1941)

Robson, A. O., Jelliffe, A. M.: Medicinal arsenic poisoning and lung cancer. Brit med. J. *2*, 207 (1963)

Rodgers, J. D. (1941): cited by Neubauer, O. (1947)

Rosen, I.: cited by Montgomery, H. (1935)

Roth, F.: Über die chronische Arsenvergiftung der Moselwinzer unter besonderer Berücksichtigung des Arsenkrebses. Z. Krebsforsch. *61*, 287 (1956)

Rousset, M.: Arsenical keratoses associated with carcinomas of the internal organs. Canad. med. J. *78*, 416 (1958)

Russel, B. F., Klaber, R. (1945): cited by Neubauer, O. (1947)

Ryan, M. L.: Arch. Derm. Syph. (Chicago) *40*, 104 (1939)

Sanderson, K. V.: Arsenic and skin cancer. Trans. St. John's Hosp. Derm. Soc. (Lond.) 115 (1963)

Schamberg, J. F. (1906): cited by Ullmann, K. (1930)

Schönhof (1923): cited by Ullmann, K. (1930)

Schwartz (1926): cited by Ullmann, K. (1930)

Schwenzner, G., Walther, H.: Multiple Basaliome auf Psoriasis vulgaris nach längerer Arsenbehandlung. Z. Hautkr. *28*, 204 (1960)

Semon, H. C. (1922): cited by Ullmann, K. (1930)

Semon, H. C.: Proc. roy. Soc. Med. *38*, 128 (1945)

Sommers, S. C., McManus, R. G.: Multiple arsenical cancers of skin and internal organs. Cancers (Philad.) *6*, 347 (1953)

Stilians, A. W.: J. cutan. Dis. *37*, 269 (1919)

Stilians, A. W.: Psoriasis, arsenical dyskeratoses, multiple epitheliomas. Arch. Derm. Syph. (Chicago) *23*, 377 (1931)

Szodoray, L. (1961): cited by Unna, P. J., Memmersheimer, A., Herzberg, J. J. (1963)

Thiers, H., Colomb, D., Moulin, G., v. Colin, L.: Le cancer cutané arsenical des viticulteurs du Beaujolais. Ann. Derm. *94*, 133 (1967)

Thiovolet, J., Bondet, P., Perrot, H., Claudy, A.: Epithéliomatose sur arsénicisme chronique thérapeutique. Bull. Soc. franç. Derm. Syph. *76*, 892 (1969)

Traub, E. F.: Basal- and prickle cell epitheliomatosis following arsenic medication. Arch. Derm. Syph. (Chicago) *36*, 198 (1937)

Traub, E. F.: Multiple epitheliomas. Arch. Derm. Syph. (Chicago) *38*, 377 (1939)

Trimble (1914): cited be Ullmann, K. (1930)

Trow, E. J.: cited by Montgomery, H. (1935)

Ullmann, K. (1898, 1917, 1922): cited by Ullmann, K. (1930)

Ullmann, K.: Neueres über das As-Carcinom. VIII Congr. int. Derm. Syph., Copenhagen, 1930

Ullmann, K.: Krebsbildung in der Gewerbemedizin und ihre Beziehung zur experimentellen Geschwulstforschung. In: Handbuch der Haut- und Geschlechtskrankheiten. Jadassohn, J. (ed.) Berlin: Springer, 1933, Vol. XII/3, pp. 551–719

Ullmann and Petschek (1906): cited by Ullmann, K. (1930)

llmann and Rona (1916): cited by Ullmann, K. (1930)

Unna, P. J., Memmersheimer, A., Herzberg, J. J.: Das Carcinom bei psoriasis vulgaris – post hoc oder proper hoc? Arch. klin. exp. Derm. *217*, 321 (1963)

Voss, F.: Multiple Arsenkarzinome. Strahlentherapie *66*, 156 (1939)

Warburg, O.: On the origin of cancer cells. Science *123*, 309 (1956)

Warren, M.: Arch. Path. (Chicago) *30*, 977 (1940)

Waugh, J. F., Scull, C. W.: Arch. Derm. Syph. (Chicago) *31*, 143 (1935)

Weidenfeld, S. (1912): cited by Ullmann, K. (1930)

White, J. C. (1885, 1899): cited by Ullmann, K. (1930)

Wright, C. S., Friedman, R. J.: Psoriasis and multiple superficial epithelioma. Arch. Derm. Syph. (Chicago) *27*, 70 (1933)

Wile, U. J. (1912): cited by Ullmann, K. (1930)

Wilhelm, L. F. X.: cited by Ayres, S. (1934)

Wilhelm, L. F. X., Goeckermann, W. H. (1940, 1943): cited by Neubauer, O. (1947)

Wiliamson, A. W. R.: Arsenical skin cancer and lung cancer. Guys Hosp. Rev. *109*, 42, (1960)

Wise, F. (1920): cited by Neubauer, O. (1947)

Yeh, S., How, S. W., Lin, C. S.: Arsenical cancer of skin. Cancer (Philad.) *21*, 312 (1968)

Zaun, H.: Iatrogene Arsenschäden unter besonderer Berücksichtigung von Hautkrebsen. Dtsch. Ärzteblatt *22*, 1217 (1965)

10

Table 1. Tumors following the use of medicinal arsenic

Patient Age	Sex	Drug and duration of treatment	Type of tumor	Reference
45	M	Arsenic (25 years)	Epithelioma	Pozzi, 1874
40	M	Arsenic (23 years)	Epithelioma	Cartaz, 1877
46	M	Fowler's solution (6 years)	Epitheliomas	White, 1885
52	M	Arsenic (long period of time)	Epitheliomas	,,
35	M	Arsenic (long period of time)	Epitheliomas	v. Hebra, 1887
34	M	Arsenic (long period of time)	Epithelioma	Hutchinson and Tay, 1888
25	F	Arsenic (long period of time)	Epithelioma	Hutchinson and Albutt, 1888
55	M	Arsenic	Epitheliomas	Hutchinson, 1888
63	M	Arsenic (30 years)	Epitheliomas	Lane, 1894
35	M	Arsenic (10 months)	Epitheliomas	Hutchinson, 1894
46	M	Arsenic (many years)	Epitheliomas	Hutchinson and Bullock, 1898
37	F	Fowler's solution (20 months)	Epitheliomas	Hutchinson, 1898
37	F	Fowler's solution (6–7 years)	Epitheliomas: basal cell carcinoma, squamous cell carcinoma	Ullmann, 1898
35	F	Arsenic (long period of time)	Epitheliomas	Hartzell and Stelwagon, 1899
–	M	Arsenic (long period of time)	Epithelioma	Hyde, 1899
–	M	Arsenic (long period of time)	Epithelioma	White, 1899
60	M	Arsenic (2 years)	Epitheliomas: squamous cell carcinomas	Crocker and Pernet, 1901
35	M	Arsenic (5 years)	Epitheliomas	Brocq, 1902
47	M	Arsenic (many years)	Epitheliomas	Darier, 1902
–	–	Arsenic	Epithelioma	Heuss, 1902
70	M	Arsenic (long period of time)	Epitheliomas	Hutchinson, 1902
62	M	Arsenic (many years)	Epithelioma	Hutchinson, 1903
55	M	Fowler's solution	Epithelioma	Hartzell, 1906
62	M	Fowler's solution (25 years)	Epitheliomas	Schamberg, 1906
42	M	Fowler's solution (3 years)	Epitheliomas: 3 squamous cell carcinomas	Ullmann and Petscheck, 1906

Table 1 (continued)

Patient Age	Sex	Drug and duration of treatment	Type of tumor	Reference
–	F	Arsenic	Breast carcinoma	Ullmann and Petscheck, 1906
–	F	Arsenic	Breast carcinoma	,,
71	M	Fowler's solution (20 years)	Epithelioma: squamous cell carcinoma	Dubreuilh, 1910
–	F	Arsenic (25 years)	Breast carcinoma	,,
56	F	Arsenic (32 years)	Epithelioma: basal cell carcinoma	Gray, 1912
38	F	Arsenic (20 years)	Epitheliomas: squamous cell carcinoma	Weidenfeld, 1912
29	M	Arsenic (2 years)	Epithelioma	Wile, 1912
50	M	Arsenic (several years)	Epitheliomas: 2 squamous cell carcinomas	Löwenberg, 1913
50	M	Arsenic (several years)	Epithelioma	,,
29	F	Arsenic (3 years)	Epitheliomas	Pye-Smith and Hardwoodnutt, 1913
65	M	Arsenic (long period of time)	Epitheliomas?	Trimble, 1914
37	M	Fowler's solution (long period of time)	Epitheliomas: basal cell carcinomas	Fordyce and McKee, 1914
56	M	Asiatic pill and Fowler's solution (3–4 years)	Epitheliomas: squamous cell carcinomas	Ullmann and Rona, 1916
34	F	Fowler's solution (2.5 years)	Epitheliomas: 1 squamous cell carcinoma 1 carcinoma of the tongue	Ullmann, 1917
60	F	Fowler's solution (30 years)	Epithelioma	Bland-Sutton, 1916
57	M	Arsenic (1 year)	Epithelioma	Stillians, 1919
–	–	Arsenic	Epithelioma	Pusey, 1919
51	M	Arsenic (25 years)	pitheliomas: squamous cell carcinomas	Rasch and Fons, 1920
–	–	Arsenic	Epithelioma	Robba, 1920
42	M	Fowler's solution (6 weeks)	Epithelioma: squamous cell carcinoma	Wise, 1920
35	M	Fowler's solution (15 years)	Epithelioma: basal cell carcinoma	Alexander, 1921
35	M	Fowler's solution (36 years)	pithelioma: squamous cell carcinoma	Hamilton, 1921
40	M	Fowler's solution (7 years)	Epithelioma: squamous cell carcinoma	Semon, 1922

Table 1 (continued)

Patient Age Sex	Drug and duration of treatment	Type of tumor	Reference
52 F	Fowler's solution (several years)	Epithelioma: squamous cell carcinoma	Ullmann, 1922
65 M	Asiatic pill (9 years)	Epitheliomas	Haxthausen, 1923
– M	Asiatic pill (22 years)	Epitheliomas: squamous cell carcinomas	Nander, 1923
42 M	Arsenic (long period of time)	Epitheliomas	Oliver, 1923
58 M	Fowler's solution (5 years)	Epithelioma	Nicolas et al., 1923
35 F	Arsenic	Epithelioma: cancroid	Schönhof, 1923
47 M	Fowler's solution, Asiatic pill	Epitheliomas: squamous cell carcinomas	Aliferis, 1924
67 M	Fowler's solution, Asiatic pill	Epitheliomas: squamous cell carcinomas	,,
– –	Arsenic (long period of time)	Epitheliomas	Barber, 1924
66 M	Arsenic	Epitheliomas: 7 squamous cell carcinomas	Gelbjerg-Hansen, 1924
45 M	Arsphenamine, arsenic applied per os and by injection (several years)	Epitheliomas: basal cell carcinomas	Levin, 1926
35 M	Asiatic pill (8 years) and injections of arsenic	Epithelioma: squamous cell carcinoma	,,
– –	Arsenic (10 years)	Epithelioma	Herxheimer, 1926
– –	Arsenic (8 years)	Epitheliomas: squamous cell carcinomas	Schwartz, 1926
62 F	Fowler's solution (a few months)	Epitheliomas: basal cell carcinomas	Finsen, 1927
46 F	Injections of arsphenamine	Fibrosarcoma at site of injection	Harbitz, 1927
20 F	Arsenic (long period of time)	Epithelioma	Haxthausen, 1928
50 M	Arsenic	Epithelioma: squamous cell carcinoma	Bruusgard, 1928
– M	Arsenic (several years)	Epitheliomas	Hofmann, 1928
57 M	Arsenic (20 years)	Epithelioma	Konrad, 1928
64 M	Arsenic (long period of time)	Epithelioma: squamous cell carcinoma	Mayer, 1928
65 F	Arsenic (20 years)	Epithelioma: squamous cell carcinoma	Fuhs, 1929

Table 1 (continued)

Patient Age	Sex	Drug and duration of treatment	Type of tumor	Reference
–	–	Fowler's solution (11 years)	Epitheliomas	Oliver and Finnerud, 1928
–	M	Arsenic (probably for 20 years)	Epithelioma	Cole and Driver, 1929
68	M	Asiatic pill	Epithelioma	Ramel, 1929
–	–	Arsenic (30 years)	Epitheliomas: 2 squamous cell carcinomas	Rasch, 1930
61	M	Fowler's solution (2 years)	Epitheliomas: squamous cell carcinomas	Ullmann, 1930
69	M	Fowler's solution	Epitheliomas	Doty, 1931
48	F	Fowler's solution (long period of time)	Epitheliomas: basal cell carcinomas	Gross, 1931
55	M	Fowler's solution (15 years)	Epitheliomas: 6 squamous cell carcinomas	Haagensen, 1931
46	M	Fowler's solution (16 years)	Epitheliomas: basal cell carcinomas	,,
31	M	Fowler's solution (6 months)	Epitheliomas: squamous cell carcinomas	,,
35	M	Fowler's solution (8 years)	Epithelioma: squamous cell carcinoma	,,
47	F	Arsenic	Melanotic sarcoma? epithelioma: squamous cell carcinoma	Milian, 1931
–	F	Arsenic	Breast carcinoma	,,
58	M	Fowler's solution (2 years)	Epitheliomas	Stilians, 1931
75	M	Fowler's solution	Epitheliomas	Anderson, 1932
70	M	Arsenic	Epithelioma: basal cell carcinoma	,,
75	F	Fowler's solution	Epitheliomas: basal cell carcinomas	,,
–	F	Arsenic	Epitheliomas	,,
50	M	Arsenic (long period of time)	Epitheliomas: squamous cell carcinoma	Fischer, 1932
–	M	Arsenic (long period of time)	Epitheliomas: basal cell carcinomas	Barber, 1933
55	F	Fowler's solution (27 years)	Epithelioma: cancroid, basal cell carcinoma	Guggenheim, 1933

Table 1 (continued)

Patient Age Sex	Drug and duration of treatment	Type of tumor	Reference
51 F	Arsenic (2 years)	Epitheliomas: basal cell carcinoma, cervical carcinoma	McCormac, 1933
45 M	Arsenic (6 weeks)	Epitheliomas: 12 basal cell and squamous cell carcinomas	Wright and Friedman, 1933
40 F	Arsenic (2 years)	Epitheliomas	,,
60 F	Arsenic (1 year)	Epitheliomas	,,
older than 60 M	Fowler's solution	Epitheliomas	Ayres, 1934
− M	Fowler's solution (7 years)	Epithelioma	Wilhelm (cited by Ayres, 1934)
55 M	Arsenic (3– 4 years) and arsphenamine	Epitheliomas: squamous cell carcinomas	Franseen and Taylor, 1934
67 F	Fowler's solution	Epitheliomas: squamous cell carcinoma, esophageal carcinoma?	,,
63 M	Fowler's solution (2 years)	Epitheliomas: squamous cell carcinomas	,,
50 F	"Dr. Greene's Nervine"	Epitheliomas: squamous cell carcinomas	,,
51 M	Arsenic (2 months)	Epithelioma: squamous cell carcinoma	,,
53 M	Arsenic (15 years)	Epitheliomas: basal cell carcinomas and squamous cell carcinomas	,,
47 M	Arsenic (long period of time)	Pancreatic carcinoma	,,
37 M	Fowler's solution (10 years)	Epithelioma: squamous cell carcinoma	,,
65 F	Arsenic (probably)	Epitheliomas: 1 basal cell carcinoma 1 squamous cell carcinoma of the vulva	,,
63 M	Fowler's solution	Epithelioma: basal cell carcinoma	,,
53 F	Arsenic (3 years)	Epitheliomas: basal cell carcinomas	,,
53 M	Fowler's solution (short period of time)	Epitheliomas	,,

Table 1 (continued)

Patient Age	Sex	Drug and duration of treatment	Type of tumor	Reference
31	M	Fowler's solution (6 months)	Epitheliomas squamous cell carcinomas	McNeer, 1934
50	M	Fowler's solution (3 weeks)	Epitheliomas: squamous cell carcinomas	,,
37	M	Arsenic	Epithelioma	,,
50	M	Asiatic pill (7 years)	Epitheliomas	Alderson, 1935
59	F	Fowler's solution (long period of time)	Epitheliomas	Ayres, 1935
–	–	Arsenic	Epithelioma: squamous cell carcinoma	Hövelborn, 1935
34	M	Fowler's solution (5 years)	Epitheliomas: squamous cell carcinomas	Montgomery, 1935
42	M	Fowler's solution (6 years)	Epitheliomas: basal cell carcinomas, squamous cell carcinomas, bronchial carcinoma	,,
63	M	"Ascato"	Epitheliomas: basal cell carcinomas squamous cell carcinomas	,,
44	F	Fowler's solution (1 year)	Epithelioma	,,
50	M	Fowler's solution	Epitheliomas	,,
–	–	Arsenic (from time to time)	Epitheliomas	Parkhurst (cited by Waugh and Scull, 1935)
–	–	Arsenic	Epitheliomas	,,
–	–	Fowler's solution (12 years)	Epithelioma: basal cell carcinoma	Robinson, 1935
40	M	Arsphenamine	Epithelioma: oral mucosa	Rosen (cited by Montgomery, 1935)
50	M	Fowler's solution (2–3 years)	Epitheliomas	Trow (cited by Montgomery, 1935)
58	F	Fowler's solution (3 years)	Epitheliomas: basal cell carcinomas	Waugh and Scull, 1935
–	F	Arsenic	Epitheliomas: superficial carcinomas and squamous cell carcinomas	Bloom, 1936
35	M	Fowler's solution (1.5 years)	Epithelioma: basal cell carcinoma	Fassrainer, 1937

Table 1 (continued)

Patient Age Sex	Drug and duration of treatment	Type of tumor	Reference
— M	"Ascato"	Epitheliomas: benign superficial tumors and squamous cell carcinomas	Anderson, 1937
— —	Arsenic	Epitheliomas: basal cell carcinomas	Ayres (cited by Anderson, 1937)
68 F	Arsenic (4 years)	Epitheliomas: basal cell carcinomas	Traub, 1937
— M	Fowler's solution (many years)	Epitheliomas	Robinson, 1937
49 M	Fowler's solution (25 years)	Epitheliomas: basal cell carcinomas	Rauschkolb, 1938
37 M	Arsenic (10 years)	Epitheliomas: basal cell carcinomas	Barber, 1939
— F	Arsenic (20 years)	Epithelioma	Haldin-Davis, 1939
62 F	Fowler's solution, Asiatic pil	Epitheliomas: basal cell carcinomas	Voss, 1939
52 M	Arsenic (10 years)	Epitheliomas	Ryan, 1939
21 M	Arsenic (6 months)	Epitheliomas: basal cell carcinoma, carcinoma of the tongue	Anderson (cited by Goeckermann and Wilhelm, 1940)
70 M	Fowler's solution (10 years)	Papilloma ureteri et vesicae	Goeckermann and Wilhelm, 1940
— —	Arsenic	Epitheliomas: squamous cell carcinomas and basal cell carcinomas	Warren, 1940
60 F	Fowler's solution	Epitheliomas: basal cell carcinomas	Applestein, 1941
— —	Arsenic	Epitheliomas: basal cell carcinomas	Madsen, 1941
— —	Arsenic	Epitheliomas: Basal cell carcinomas, Bowen's disease	,,
49 M	Fowler's solution	Epitheliomas: basal cell carcinomas, squamous cell carcinomas	Montgomery and Waisman, 1941
69 F	Fowler's solution (6 weeks)	Epitheliomas: squamous cell carcinomas	,,
— —	Arsenic cure	Epitheliomas: basal cell carcinomas	Piper, 1941

17

Table 1 (continued)

Patient Age	Sex	Drug and duration of treatment	Type of tumor	Reference
–	–	Arsenic?	Epitheliomas: squamous cell carcinomas	Robinson (cited by Rodgers, 1941)
55	M	Fowler's solution	Epitheliomas: basal cell carcinomas, squamous cell carcinomas	Rodgers, 1941
–	–	Arsenic	Epitheliomas: squamous cell carcinomas	Hohmann, 1942
–	–	Arsenic	Epitheliomas: basal cell carcinomas	,,
42	F	Fowler's solution	Epitheliomas: Bowen's disease	Montgomery, 1942
45	M	Arsenic (3.5 years)	Epitheliomas: squamous cell carcinomas	Peck, 1942
47	F	Arsenic (injections)	Epitheliomas: basal cell carcinomas	Anderson, 1943a
50	M	Arsenic	Epitheliomas	Anderson, 1943b
56	M	Asiatic pill (5 years)	Epitheliomas: (some) squamous cell carcinomas	Ayres, 1943
42	F	Arsenic	Epitheliomas	Ebert and Otsuka, 1943
61	F	Arsenic (5 years)	Epitheliomas	Laymon, 1943
72	M	Arsenic (drops and injections)	Epitheliomas	Wilhelm and Goeckermann, 1943
46	M	Black pills (6 months)	Epitheliomas: basal cell carcinomas; esophageal carcinoma, squamous cell carcinoma	Barney, 1944
61	F	Arsenic (orally and intravenously)	Epitheliomas	Hall, 1944
–	–	Arsenic	Epitheliomas	Dowling, 1945
53	F	Fowler's solution (13 years)	Epitheliomas: squamous cell carcinomas	Russel and Klaber, 1945
44	F	Fowler's solution	Epitheliomas	–
–	M	Fowler's solution	Bronchial carcinoma	Semon, 1945
–	–	Arsenic (14 years)	Epithelioma: squamous cell carcinoma	,,
–	–	Fowler's solution	Epitheliomas: basal cell carcinomas, squamous cell carcinomas	Degos et al., 1950

Table 1 (continued)

Patient Age Sex	Drug and duration of treatment	Type of tumor	Reference
50 M	Fowler's solution (3 years)	Epitheliomas: basal cell carcinomas, squamous cell carcinomas	Arhelger and Kremen, 1951
51 F	Fowler's solution (6 months)	Epitheliomas: squamous cell carcinomas	,,
66 M	Fowler's solution (1 year)	Epithelioma: squamous cell carcinoma	,,
78 M	Fowler's solution (many years)	Epithelioma: squamous cell carcinoma	,,
70 F	Fowler's solution (32 years)	Epitheliomas: basal cell carcinomas	,,
50 M	Fowler's solution (13 years)	Epitheliomas	,,
52 F	Fowler's solution (3 months)	Epithelioma: basal cell carcinoma	,,
69 F	Fowler's solution (1 month)	Epithelioma: squamous cell carcinoma	,,
59 M	Fowler's solution	Epithelioma: basal cell carcinoma	,,
– –	Arsenic	Epithelioma: basal cell carcinoma	Amersbach, 1952
43 M	Fowler's solution (12 years)	Epitheliomas: squamous cell carcinoma, basal cell and squamous cell carcinoma, basal cell carcinoma, Bowen's disease; undifferentiated carcinoma of the bladder or lower ureter; adenocarcinoma of the prostate	Sommers and McManus, 1953
45 F	Fowler's solution (1 year)	Epitheliomas: basal cell carcinomas; adenocarcinoma of the colon	,,
46 M	Arsenic (injections)	Epithelioma: squamous cell carcinoma; undifferentiated sarcoma of the thorax; papillary kidney carcinoma	,,
54 M	Arsenic	Epithelioma: squamous cell carcinoma; bronchial carcinoma: squamous cell carcinoma	,,

Table 1 (continued)

Patient Age Sex	Drug and duration of treatment	Type of tumor	Reference
— M	Arsenic	Epitheliomas: basal cell carcinomas, squamous cell carcinomas; Bowen's disease, basal cell and squamous cell carcinoma	„
— M	Arsenic	Epitheliomas: basal cell carcinomas, hidradenoid	„
66 M	Arsenic	Epitheliomas: basal cell carcinomas	„
54 M	Arsenic	Epitheliomas: basal cell carcinomas, squamous cell carcinomas	Sommers and McManus, 1953
51 M	Arsenic	Epitheliomas: basal cell and squamous cell carcinomas, basal cell carcinomas	„
60 F	Arsenic	Epitheliomas: basal cell carcinomas, squamous cell carcinomas	„
30 M	Arsenic	Epitheliomas: basal cell carcinomas	„
47 F	Arsenic	Epithelioma: basal cell carcinoma with keratin and hidranoid foci	„
56 M	Arsenic (26 years; 15 arsenic cures)	Bronchial carcinoma	Braun, 1958
51 M	Fowler's solution (20 years)	Anaplastic cholangioma or hemangioendothelioma of the liver	Rousset, 1958
— —	Arsenic	Epithelioma: Bowen's disease; squamous cell carcinoma of the portio; lymphangiosis carcinomatosa of the lung	Doepfmer, 1959/60
— —	Fowler's solution (4 cures)	Epitheliomas: squamous cell carcinomas, Bowen's disease	Kleine-Natrop, 1959

20

Table 1 (continued)

Patient Age Sex	Drug and duration of treatment	Type of tumor	Reference
53 M	Arsenic	Epitheliomas: basal cell carcinomas	Schwenzner and Walther, 1960
50 M	Arsenic (18 months)	Epitheliomas: Bowen's disease; bronchial carcinoma	Williamson, 1960
58 M	Arsenic	Epitheliomas: squamous cell carcinoma; bronchial carcinoma: squamous cell carcinoma	,,
– –	Arsenic	Epitheliomas: basal cell carcinomas	Szodoray, 1961
58 M	Arsenic ("in litres")	Epitheliomas: squamous cell carcinomas, adenocarcinomas; bronchial carcinoma, squamous cell carcinoma, adenocarcinoma (metastasis)	Berlin and Tager, 1962
31 F	Arsenic (serveral years)	Epitheliomas: intraepicutaneous carcinoma; bronchial carcinoma: undifferentiated carcinoma	Robson and Jelliffe, 1963
45 F	Fowler's solution (15 years)	Bronchial carcinoma: undifferentiated carcinoma	,,
54 M	Fowler's solution (4 years)	Epitheliomas: bronchial carcinoma: hardly differentiated carcinoma	,,
47 F	Arsenic	Epitheliomas: intraepicutaneous carcinoma; bronchial carcinoma	,,
65 M	Fowler's solution	Epithelioma: intraepidermic carcinoma;	,,

21

Table 1 (continued)

Patient Age Sex	Drug and duration of treatment	Type of tumor	Reference
60 F	Fowler's solution (4 years)	Bronchial carcinoma: hardly differentiated carcinoma	,,
22 M	Arsenic (30 months)	Epitheliomas: intraepidermic carcinomas	Sanderson, 1963
59 M	Fowler's solution (several years)	Epitheliomas: basal cell carcinomas, intraepidermic carcinomas; bronchial carcinoma: anaplastic carcinoma	,,
37 M	Arsenic liquor (1 year)	Epitheliomas: intraepidermic carcinomas	,,
37 F	Arsenic	Epitheliomas: squamous cell carcinomas	,,
38 F	Arsenic?	Epithelioma: basal cell carcinoma	,,
36 M	Arsenic	Epithelioma: basal cell carcinoma	,,
43 M	Arsenic liquor (2 years)	Epitheliomas: basal cell carcinomas	,,
48 M	Arsenic?	Epitheliomas: basal cell carcinomas	,,
48 F	Arsenic	Epitheliomas: basal cell carcinomas	,,
54 M	Arsenic (several months)	Epitheliomas: basal cell carcinomas	,,
60 M	Fowler's solution (10 cures)	Epitheliomas: squamous cell carcinomas	Unna et al., 1963
67 M	Arsenic (2 cures)	Epitheliomas: squamous cell carcinomas	,,
60 —	Fowler's solution and arsenettes	Epitheliomas: basal cell carcinomas, squamous cell carcinomas	,,
50 M	Fowler's solution (3 months) and elpsoral (6 packings)	Epithelioma: squamous cell carcinoma	,,

Table 1 (continued)

Patient Age Sex	Drug and duration of treatment	Type of tumor	Reference
21 M	Fowler's solution (3 years; total dose: 7.8 g), As 203	Epitheliomas: Squamous cell carcinomas	Minkowitz, 1964
52 M	Arsenic (drops; 5 bottles)	Epitheliomas: basal cell carcinomas	Zaun, 1965
34 M	Acidic arsenic (total dose: 1–1.5 g)	Epithelioma: squamous cell carcinoma	"
42 M	Fowler's solution, Arsenic (total dose: 2000 ml)	Epitheliomas: basal cell carcinomas, Bowen's disease, squamous cell carcinomas	Meyhöfer and Knoth, 1966
53 M	Arsenic (several cures)	Reticulosarcoma of the glans penis	Knoth, 1966
61 F	Arsenic (longer period of time)	Epitheliomas: Bowen's disease, breast carcinoma	"
49 M	Fowler's solution (17 years; total dose: 53 g), arsenic	Hemangioendothelial sarcoma	Regelson et al., 1968
56 F	Arsenic (30 years; total dose: 21 g)	Epitheliomas: 1 Bowen's carcinoma in situ	Novey and Martel, 1969
72 F	Fowler's solution (43 years)	Epitheliomas: Bowen's disease, Bowen's carcinoma	Thiovolet et al., 1969
26 F	Fowler's solution	Epithelioma: basal cell carcinoma	Bartak and Kejda, 1972

23

Table 2. Basic diseases of patients with arsenic-related tumors

A. Dermatoses	Number of cases
Psoriasis vulgaris	115
Eczema, dermatitis	7
Acne	7
Dermatitis herpetiformis	4
Pemphigus	3
Acuminatus pityriasis rubra pilaris	2
Cosmetic reasons	1
Undifferentiated skin diseases	3
Total number	142

B. Extracutaneous diseases	Number of cases
Epilepsy	11
Choréa	6
"Attacks"	2
"Nerves"	4
Asthma, chronic bronchitis	9
Syphilis	4
Ulcer of the mouth (syphilis?)	1
Rheumatic fever	4
Anemia	6
Pernicious anemia	1
Hemophilia	1
Malaria prophylaxis	1
General debility	2
Pulmonary tuberculosis	1
Pulmonary apicitis	1
Gastric disorders	1
Cervical nodes	1
Total number	56

Table 3. Localizations of arsenic-related epitheliomas (185 cases)

Localization	Number of cases
Trunk (including shoulder, axilla, breast, hip, and sacral region)	106
Hand (including wrist joint and fingers)	68
Legs	26
Arms	18
Neck, back of the neck	13
Penis, scrotum	12
Anus, perineum	7
Vulva	1
Foot (including heel and toes)	32
Head	25
Total number of cases	308[a]

[a] The total number of cases amounts to more than 185 because frequently several tumor localizations were described.

Table 4. Histologic diagnoses of arsenic-related skin tumors (143 cases)

Histologic diagnosis	Number of cases
Squamous cell carcinoma	58
Basal cell carcinoma	48
Basal cell and squamous cell carcinoma	15
Combined basal cell and squamous cell carcinoma	5
Cancroid	2
Intraepidermic epithelioma	3
Intraepidermic carcinoma	4
Fibrosarcoma	1
Melanotic sarcoma	1
Bowen's carcinoma	2
Bowen's disease	10
Total number of cases	149[a]

[a] The total number of cases amounts to more than 143 since besides Bowen's disease further epitheliomas were developed.

25

Table 5. Arsenic-related visceral tumors occurring besides epitheliomas (19 cases)

Type of tumor	Number of cases
Bronchial carcinoma	9
Esophageal carcinoma	2
Cervical carcinoma	2
Carcinoma of the tongue	2
Carcinoma of the breast	1
Colonic carcinoma	1
Carcinoma of the bladder	1
Carcinoma of the prostate	1
Papillary kidney carcinoma	1
Undifferentiated sarcoma of the thorax	1
Total number of cases	21[a]

[a] The total number of cases amounts to more than 19 because in two cases more than one tumor was observed.

Table 6. Arsenic-related visceral tumors occurring without coexistent epitheliomas (11 cases)

Type of tumor	Number of cases
Bronchial carcinoma	4
Carcinoma of the breast	2
Pancreatic carcinoma	1
Papilloma of the bladder and the ureter	1
Hemangioendothelioma or anaplastic cholangioma of the liver	1
Hemangioendothelial sarcoma of the liver	1
Reticulosarcoma of the glans penis	1
Total number of cases	11

Chloramphenicol

Chloramphenicol has been used as an antibiotic since 1949. It is inactivated in the liver by glucuronidation and rapidly excreted in the urine (Goodman and Gilman, 1970).

By 1950, Rich reported the first case of aplastic anemia associated with chloramphenicol medication. Its toxicity to the hematopoietic system, resulting in leukopenia, thrombocytopenia, pancytopenia, and aplastic anemia, has since been verified in numerous publications (Kähler, 1962; Best, 1967; Polak et al., 1972).

Lebon and Messerschmitt (1955) drew attention to a possible causal role of chloramphenicol in the formation of human leukemia.

In a survey of the literature we found 17 cases of acute leukemia, possibly associated with the use of chloramphenicol, which are shown in Table 7. One peak in incidence was observed in childhood and a second between the 50th and 70th year of age. Chloramphenicol had been prescribed for tonsillitis or infections of the upper respiratory tract (four cases), cystitis and pyelonephritis (three cases), prostatitis (one case), postoperative prophylaxis (one case), frequent colds (two cases), whooping cough (two cases), pneumonia accompanied by cystitis, and septicemia (one case each). It had been administered over periods of 3 days to 180 months. Total doses ranged from 5–230 g. Latent periods ranged from 30 days[2] to 180 months (average latent period: 41 months).

The following types of leukemia were formed: ten acute myeloid leukemias, four acute stem cell leukemias, two acute lymphatic leukemias, and one acute myelomonocytic leukemia.

Four patients had also received other myelotoxic drugs.

Previous bone marrow injuries were present in 12 cases.

In eight cases the course of the disease could be traced: two children were without relapse after 33 and 36 months of cytostatic treatment, respectively; one child died after 30 months and a second as well as three adult patients died shortly after the diagnosis; one adult died after 14 months.

The myelotoxic part of the chloramphenicol molecule is not known. The effect was initially attributed to the nitro group of the benzene ring (Smadel, 1949), a view that had to be revised after Suhrland and Weisberger (1962) discovered that after substitution by other groups the toxicity is not diminished but increased.

In chloramphenicol-induced lesions of bone marrow we can distinguish between reversible and irreversible types. The reversible lesions, according to Yunis and Bloomberg (1964), are dose-related and occur during treatment. They are characterized by anemia, reticulocytopenia, disorders in iron metabolism, and vacuolization of erythroblasts and the first stages of granulocytes. The irreversible lesions are not dose-related

---

[2] In this case, because of the short latency period, a causal relation between chloramphenicol therapy and leukemia is very unlikely

and develop weeks to months after treatment. They occur in the form of severe hypoplasia or aplasia of the bone marrow.

Chloramphenicol attacks the mRNA and thus inhibits protein synthesis (Weisberger, 1969). This would explain its immunosuppressive activity (Weisberger and Daniel, 1969). The leukemogenic action of chloramphenicol has been related to chromosomal vacuolization of bone marrow (Brauer and Dameshek, 1967; Goh, 1971). Castoldi and Mitus (1968) observed chromosomal vacuolizations in a chloramphenicol-treated patient, and Goh (1971) described a case of simultaneous chloramphenicol therapy, acute leukemia, and chromosomal vacuolization. He pointed out that those vacuolizations were not observed in the smear of bone marrow of healthy individuals or in that from patients with other types of leukemia.

The myelotoxicity of chloramphenicol is thought to be responsible for the formation of acute leukemia. This view is supported by the formation of acute leukemia in the course of aplastic anemia. These symptoms may therefore occur independently (Brauer and Dameshek, 1967). Aplastic anemia could also be interpreted as a preleukemic stage or an atypical leukemia (Block et al., 1953; Blair et al., 1966).

Chromosomal aberrations were also observed in benzene-induced leukemias (Hartwich et al., 1969). A relation between chromosomal aberrations and acute leukemias in Down's syndrome (trisomy 21) was established by Krivit and Good (1964), who observed a 3- to 15-fold increase in the rate of leukemia in mongoloids. Gadner et al. (1973) proposed that chloramphenicol as well as bone marrow hyperplasia might favor infection by an oncogenic virus (?).

The following mechanisms of leukemia formation following the use of chlhave been considered:

1. Interference with the immune response
2. Favoring infection by an oncogenic virus
3. Production of chromosomal injuries
4. Production of bone marrow lesions.

In 13 of the cases reported a causal relationship between chloramphenicol medication and acute leukemia is possible, whereas in the remaining four patients bone marrow hypoplasia or acute leukemia may have been present before chloramphenicol application was started. The peaks in incidence, the first in childhood and the second between the 50th and 70th year of age, agree with those observed in other leukemias (Begemann, 197). As is generally the case, acute myeloid leukemias occurred exclusively in adult patients, whereas in children acute nonmyeloid leukemias predominated.

28

References

Begemann, H.: Klinische Hämatologie. Stuttgart: Georg Thieme 1970

Best, W. R.: Chloramphenicol-associated blood dyscrasias. J. Amer. med Ass. *201*, 99 (1967)

Blair, T. R., Bayrd, E. D., Pease, G. L.: Atypical leukemia. J. Amer. med. Ass. *198*, 21 (1966)

Block, M., Jacobson, L. O., Bethard, W. F.: Preleukemic acute human leukemia. J. Amer. med. Ass. *152*, 1018 (1953)

Brauer, M. J., Dameshek, W.: Hypoplastic anemia and myeloblastic leukemia following chloramphenicol therapy. Report of three cases. New Engl. J. Med. *277*, 1003 (1967)

Castoldi, G., Mitus, W. J. (1968): cited by Goh, K. O. (1971)

Cohen,T., Creger, W. P.: Acute myeloid leukemia following seven years of aplastic anemia induced by chloramphenicol. Amer. J. Med. *43*, 762 (1967)

Fraumeni, J. F.: Bone marrow depression induced by chloramphenicol or phenylbutazoe. J. Amer. med. Ass. *201*, 828 (1967)

Gadner, H., Gethmann, U., Jessenberger, K., Riehm, H.: Acute Leukämie nach Chloramphenicol-Exposition? Ein kasuistischer Beitrag mit Literaturübersicht. Mschr. Kinderheilk. *121*, 590 (1973)

Goh, K. O.: Chloramphenicol, acute leukemia and chromosomal vacuolizations. Sth. me. J. *64*, 815 (1971)

Goodman, L. S., Gilman, A.: The pharmacological basis of therapeutics. New York: The Macmillan Co., 1970

Hartwich, G., Schwanitz, G., Becker, J.: Chromosomenaberrationen bei einer Benzolleukämie. Dtsch. med. Wschr. *94*, 1229 (1969)

Humphries, K. R.: Acute myelomonocytic leukaemia following chloramphenicol therapy. N. Z. med. J. *68*, 248 (1968)

Kähler, H. J.: Kritische Beurteilung der Bluterkrankungen nach Anwendung von Chloramphenicol. Stuttgart: Wissenschaftliche Verlagsgesellschaft 1962

Krivit, W., Good, R. A.: Simultaneous occurrence of mongolism and leukemia. J. Pediat. *65*, 303 (1964)

Lebon, J., Messerschmitt, J.: Myélose aplasique d'origine médicamenteuse myéloblastose aigue terminale réflexions pathogéniques. Sang *26*, 799 (1955)

Mukherji, P. S.: Acute myeloblastic leukaemia following chloramphenicol treatment. Brit. Med. J. *1*, 1286 (1957)

Polak, B. C. P., Wesseling, H., Schut, D., Herxheimer, A., Meyler, L.: Blood dyscrasias attributed to chloramphenicol. Acta Med. Scand. *192*, 409 (1972)

Rich, M. L.: Fatal case of aplastic anemia following chloramphenicol (chloromycetin) therapy. Ann. Intern. Med. *33*, 1459 (1950)

Smadel, J. E.: Chloramphenicol (chloromycetin) in the treatment of infectious disease. Amer. J. Med. *7*, 671 (1949)

Suhrland, L. G., Weisberger, A. S.: Hematological toxicity of a chloramphenicol analogue. Amer. J. Med. Sci. *244*, 54 (1962)

Weisberger, A. S.: Mechanisms of action of chloramphenicol. J. Amer. Med. Ass. *209*, 97 (1969)

Weisberger, A. S., Daniel, T. M.: Suppression of antibody synthesis by chloramphenicol analogs. Proc. Soc. Exp. Biol. (N. Y.) *131*, 570 (1969)

Yunis, A. A., Bloomberg, G. R.: Chloramphenicol toxicity: Clinical features and pathogenesis. Progr. Hematol. *4*, 138 (1964)

Table 7. Human leukemias associated with the use of chloramphenicol

Patient Age	Sex	Type of leukemia	Reference
5	M	Acute myeloid leukemia	Lebon and Messerschmidt, 1955
63	M	Acute myeloid leukemia	Mukherji, 1957
47	F	Acute myeloid leukemia	Cohen and Greger, 1967
67	F	Acute myeloid leukemia	Fraumeni, 1967
2	F	Acute lymphatic leukemia	„
14	M	Acute stem cell leukemia	„
17	F	Acute stem cell leukemia	„
60	M	Acute myeloid leukemia	„
56	F	Acute myeloid leukemia	„
38	F	Acute myeloid leukemia	Brauer and Dameshek, 1967
57	F	Acute myeloid leukemia	„
61	F	Acute myeloid leukemia	„
80	M	Acute myelomonocytic leukemia	Humphries, 1968
58	M	Acute myeloid leukemia	Goh, 1971
10	M	Acute leukemia	Gadner et al., 1973
5	M	Acute leukemia	„
5	F	Acute leukemia	„

Cytostatics

In recent years the number of publications containing reports of carcinogenic effects of various cytostatic cancer chemotherapeutic agents in man has increased (reviews by Rieche, 1969; Rupprecht and Blessing, 1973; Karchner et al., 1974; Hunstein and von Rehn, 1975; Gärtner et al., 1975; IARC Monograph Series, 1975; Schmalzl et al., 1975; Sotrel et al., 1976; Sieber and Adamson, 1976; Labedzki et al., 1976).

Alkylating agents and antimetabolites disturb the synthesis of DNA and RNA or alter their structures. They are thus theorectically capable of producing teratogenic, mutagenic, and carcinogenic effects. Carcinogenicity of alkylating cytostatic agents was established by several animal experiments (even in primates) (Shimkin et al., 1966; Schmähl, 1967, 1970; Schmähl and Osswald, 1970; Weisburger et al., 1975; Philips et al., 1975; Maltoni and Chieco, 1975).

Publications containing reports of possible carcinogenic effects of cytostatic drugs in children are particularly remarkable in this respect (Rupprecht and Blessing, 1973; Gutjahr and Spranger, 1975). Since in several types of childhood cancer (e. g. acute leukemias) cytostatic treatment provides a relatively favorable prognosis, thorough follow-up observation is particularly important in these cases in order to detect possible late effects of the treatment. In Germany such efforts are in progress.

From observations in the field of occupational medicine it is well known that mustard gas can induce cancer (Wada et al., 1968; Weiss and Weiss, 1975); it is therefore probable that overdosage of alkylating cytostatics, often necessary to effect tumor inhibition, causes a chronic alkylant (= mustard gas) poisoning that may be responsible for the formation of a second tumor. However, not all alkylating substances exhibit carcinogenic effects (Schmähl and Habs, 1976).

In cancer patients it is often difficult to prove carcinogenic effects of cytostatic treatment because frequently the first tumor leads to death before the second (cytostatically-induced) tumor is observed (Hartwich and Butzler, 1971). Moreover, cancer therapy is often a polytherapy, including irradiation, so that it may be difficult to associate a second tumor with any particular agent. Nevertheless, numerous cases of second tumor associated with the use of alkylating cytostatics have been published, some of which will be considered here.

Evans et al. (1951) described the case of a 48-year-old patient with chronic myeloid leukemia who had been treated intermittently with urethan (daily dose: 1–3 g) for 15 months. He died with a metastatic adenocarcinoma of the liver that occurred during the treatment (for references on experimental urethan carcinogenesis see Schmähl, 1970). The case reports described here are of historic rather than of actual interest.

Kahlau (1954) described two patients with primary bronchial carcinoma following long-term treatment of lymphatic leukemia with urethan (dose: up to 116 g in the course of 3 years).

Heckner (1959) reported the case of a 30-year-old man with chronic myeloid leukemia who developed a reticulum cell sarcoma after 9 months' treatment with an uncommonly large dose of Myleran.

Angus and Gunz (1963) described a 34-year-old patient with chronic granulocytic leukemia who developed a pancreatic carcinoma after irradiation of the spleen and application of radioactive phosphorus, 380 g busulphan, and 6-mercaptopurine.

Nelson and Andrews (1964) observed a carcinoma of the breast in a 57-year-old woman with chronic myeloid leukemia who had been treated with busulphan for 5 years.

Terbrüggen (1965) described three patients with breast cancer, bronchial carcinoma, and adenocarcinoma of the sigma, who developed reticuloses 10 months to 4.5 years after discontinuation of a post-operative Trenimon treatment.

31

Storti et al. (1967) observed lymphosarcomatoses and reticuloses following treatment of myeloid leukemias with urethan, trisethylene-melamine, 6-mercaptopurine, cyclophosphamide, and busulphan.

Wildhack (1967) reported the case of a 75-year-old man who had been irradiated and treated with 79.4 mg Trenimon after removal of a myosarcoma of the right groin. Eighteen months after discontinuation of Trenimon monocytic leukosis and IgG-paraproteinemia occurred.

Klemm et al. (1968) described two lymphoreticular sarcomas, one reticulum cell sarcoma, and one transitional plasmocytoma following treatment with cyclophosphamide-melphalan and cyclophosphamide-chlorambucil.

Michot (1968), in a 65-year-old patient with multiple myeloma, observed a remission of the paraprotein-forming plasmocytoma 3 years after the start of treatment with cyclophosphamide and Prednison. After the therapy was continued with melphalan, acute reproliferations without formation of paraprotein occurred that were followed by angiosarcomas of the skin and subcutis.

A 72-year-old patient with chronic granulocytic leukemia died with bronchial sarcoma, diffuse interstitial lung fibroses, and atypical alveolar cell hyperplasias following 5 years' treatment with busulphan. The bronchial carcinoma and atypical alveolar changes were considered as different stages of the same pathological process (Min and Györkey, 1968).

Reis et al. (1969) reported two patients with chronic lymphadenosis and plasmocytoma who developed a mammary carcinoma and a carcinoma of the urinary bladder respectively after 5 and 9 years' treatment with trisethylene-melamine-leukeran and cyclophosphamide.

Rieche (1969) observed a gallbladder carcinoma that occurred after radical operation of a metastastic mammary carcinoma and 2 and 15 months' postoperative treatment with cyclophosphamide and Trenimon.

A 65-year-old patient with cylindric cell adenoma, after lobectomy of the upper lobe of the lung and 10 days' treatment with 6-azaridine, developed a myosarcoma of the thigh 8 months after the start of treatment (Rieche, 1969).

Kyle et al. (1970) observed three patients with plasmocytoma and one with "plasma cell dyscrasia," all of whom developed acute myelomonocytic leukemias after 30–57 months' treatment with melphalan.

Stavem and Harboe (1971) reported on a cold agglutination syndrome, in which the cold agglutination titre receded after melphalan therapy. Twenty-seven months after the start of treatment they diagnosed an acute erythroleukemia.

Craig and Rosenberg (1971) observed an increased cancer incidence in Methotrexate-treated patients with psoriasis. This finding was, however, contradicted by recent results of Baihin et al. (1975).

In Table 8 we have listed 10 cases of cyclophosphamide-related bladder carcinoma, for which the total applied doses could be determined. These ranged from 100–250 g. The mean latent period of the bladder tumors was 7 years.

The published cases are probably only a fraction of the real number of cyclophosphamide-related bladder carcinomas. This became more apparent in a personal discussion that one of us (D. Schmähl) had with an American pathologist. The American colleague mentioned that among his patients there were also "three or four" second tumors following cyclophosphamide therapy. However, he felt that publication of these cases was not necessary because the iatrogenic carcinogenicity of cyclophosphamide is generally known.

In most of the present cases the cytostatics were applied for the treatment of hemoblastoses. Hartwich and Butzler (1971) found no increased rate of second tumors in a study of 1057 patients with malignant system diseases, most of whom were treated with cytostatics. These authors, however, did not exclude the formation of second tumors in these patients and supposed that the hemoblastoses reduced the life expectancies and thus hindered clinical appearance of these tumors.

The latent periods of the second tumors were relatively short. Possibly they were reduced by the immunodepressive action of the cytostatic compounds. This hypothesis, however, could not be confirmed experimentally (Schmähl, 1971; Schmähl et al., 1974; Scherf and Schmähl, 1975).

A clear instance of cytostatically-induced carcinogenesis, which could have been avoided, is that caused by chlornaphazine (2-(bis-)chloro-ethylamino-naphthaline).

Chlornaphazine

The appearance of this substance for the chemotherapy of cancer indicates that the manufacturers had no understanding of chemical carcinogenesis. From the chemical formula of this compound it is clearly obvious that after biotransformation of the alkyl rest, which is already carcinogenic in itself, the highly carcinogenic β-naphthylamine is formed. Carcinogenic effects of some "aniline derivatives" have been known since Rehn (1895) described an increased rate of bladder cancer in persons working with dyes.

Chlornaphazine was introduced in 1948 and used in Scandinavia and Italy, mainly for the treatment of polycythemia vera and Hodgkin's disease. Chievitz and Thiede (1962) refered to general side-effects of this drug, which include dyspepsia and cystitis. Besides these they described two carcinomas of the urinary bladder and kidney carcinoma following therapy with chlornaphazine. The drug was withdrawn from the market in 1963.

Decomposition of chlornaphazine leads to the formation of β-naphthylamine, which is oxidized in the liver to form β-naphthylhydroxylamine. It is then conjugated with glucuronic acid and excreted in the urine. With high glucuronidase activity, which is present, i. e., in man and dogs, β-naphthylhydroxylamine is hydrolyzed and released.

In Table 9 we compiled 15 examples of a causal relation between chlornaphazine medication and carcinoma of the urinary bladder. The cases include seven men and eight women, aged 30–71 years (average age: 57 years). Daily doses of chlornaphazine ranged from 100 to 400 mg; total doses were between 4 and 350 g (average total dose: 150 g). Duration of treatment ranged from 9 months to 11 years, short interruptions not considered (average duration of treatment: 6 years).

The latent periods of these tumors, ranging from 2.5 to 10 years, were strikingly short (average latent period: 6 years). In five cases the tumor localizations were given; these tumors had developed in the posterior wall of the urinary bladder in the surroundings of the ureteric ostia. The following histologic diagnoses were described: "carcinoma" (five cases), "solid carcinoma" (seven cases), "papilloma" (two cases), "papillary carcinoma" (one case). One patient had metastases in the suprarenal gland and bones.

β-naphthylhydroxylamine is believed to be the direct cause of chlornaphazine carcinogenicity (Videback, 1964). The carcinogenicity of this substance or its metabolites was demonstrated in animal experiments. Hueper et al. (1948) induced bladder carcinomas in dogs using β-naphthylamine. Scott and Boyd (1953) found malignant changes only in those regions of the bladder that were in direct contact with the urine. This finding supports the importance of the glucuronidase that releases the ultimate carcinogen.

According to Thiede and Christensen (1969) and Clayson (1972) other potential carcinogens, such as 32 p or busulphan, both used additionally for the treatment of polycythemia vera, were probably not involved in the induction of these tumors.

Table 8. Cyclophosphamide-related bladder tumors in which the total doses of cyclophosphamide could be determined

Patient Age	Sex	Primary tumor	Cytostatic agent	Total dose (g)	Latency period (years)	Second tumor	Reference
47	M	Myeloma	Cyclophosphamide	295	5.4	Carcinoma of the bladder	Wall and Clausen, 1975
27	M	Hodgkin's disease	Cyclophosphamide, chlorambucil	147 23	13.5	Carcinoma of the bladder	,,
60	M	Myeloma	Cyclophosphamide	192	4	Carcinoma of the bladder, leukemia	,,
57	M	Myeloma	Cyclophosphamide	252	5.8	Carcinoma of the bladder	,,
63	M	Myeloma	Cyclophosphamide	114	3.1	Carcinoma of the bladder	,,
4	M	Hodgkin's disease	Cyclophosphamide, Mitarson	~ 3 ~ 80	10	Sarcoma of the bladder	Rupprecht et al., 1973
39	M	Hodgkin's disease	Cyclophosphamide	~ 85	10	Carcinoma of the bladder	Worth, 1971
67	M	Lymphosarcoma	Cyclophosphamide	~ 150	4	Carcinoma of the bladder	,,
73	M	Myeloma	Cyclophosphamide	~ 150	5	Carcinoma of the bladder	Dale and Smith, 1974
59	M	Waldenström's disease	Chlorambucil, Cyclophosphamide	~ 18	2	Carcinoma of the bladder	
29	F	Hodgkin's disease	Chlornaphazine cyclophosphamide, Natulan	~ 85 ~ 45 ~ 10	11	Carcinoma of the bladder	Laursen, 1970
58	F	Hodgkin's disease	Chlornaphazine, cyclophosphamide, Dichloren	78 ~ 100 0.024	10	Carcinoma of the bladder	,,

Table 9. Carcinomas of the urinary bladder following chlornaphazine therapy

Patient Age	Sex	Type and localization of tumor	Reference
68	F	Carcinoma of the urinary bladder	Videbaek, 1964
30	M	Carcinoma of the urinary bladder	„
45	F	Carcinoma of the urinary bladder	„
61	M	Carcinoma of the urinary bladder	Thiede and Christensen, 1969
54	F	Carcinoma of the urinary bladder	„
71	F	Carcinoma of the urinary bladder	„
64	M	Carcinoma of the urinary bladder	„
62	M	Carcinoma of the urinary bladder	„
67	F	Carcinoma of the urinary bladder	„
65	M	Carcinoma of the urinary bladder	„
40	M	Carcinoma of the urinary bladder	„
69	F	Carcinoma of the urinary bladder	„
60	M	Carcinoma of the urinary bladder	„
68	F	Carcinoma of the urinary bladder	Laursen, 1970
40	F	Carcinoma of the urinary bladder	„

Thiede and Christensen (1969), in a study of sixty-one chlornaphazine-treated patients with polycythemia vera, found 10 tumors of the urinary bladder and abnormal cells in the urine of five patients. Comparison of these tumors with those induced by other agents showed no difference in the age incidence. The peak in incidence, between the 60th and 70th year of age, corresponds to general observations. The sex distribution of the bladder tumors was not in accordance with that of other bladder tumors. The normal sex ratio is 3:1–5:1 in favor of men, whereas in the present cases it was 7:8 in favor of women. This difference may be attributed to the small number of cases and the differences in the basic disease. It is remarkable that metastases were described in only one case, whereas normally they are observed in about 50% of the cases. It is, however, assumed that metastases were also formed in most of the present cases (nine patients died within the first 2 years after the diagnosis). The poor prognosis corresponds to the mean life expectancy of patients with other bladder tumors (12.5 months).

Finally, we would like to emphasize that the experimental study of possible carcinogenic effects of combination schemes used in cancer chemotherapy should be intensified. It was found that patients with Hodgkin's disease treated with a combination of nitrogen mustard, vincristine, procarbazine, and prednisone (MOPP) had an almost 4-fold increase in the risk of cancer over that calculated for a comparable normal population (DeVita et al., 1973)[3].

References

Angus, B., Gunz, F. W.: Chronic granulocytic leukemia and cancer. Blood *22*, 88 (1963).

Aron, E.: Le traitement médical de la maladie de Dupuytren par un agent cytostatique (méthyl-hydrazine). Presse méd. *76*, 1956 (1968)

Baihin, P. L., Tindall, J. P., Roenigk, H. H., Hogan, M. D.: Is methotrexate therapy for psoriasis carcinogenic? J. A. M. A. *232*, 359 (1975)

Chievitz, E., Thiede, T.: Complications and causes of death in polycythaemia vera. Acta med. scand. *172*, 513 (1962)

Clayson, D. B.: Carcinogenic hazards due to drugs. In: Drug-Induced Diseases. Meyler, L., Peck, H. M. (eds.) Amsterdam: Excerpta Medica, 1972, Vol. IV, pp. 91–109

Craig, S. R., Rosenberg, E. W.: Methotrexate-induced carcinoma? Arch. Derm. *103*, 505 (1971)

Dale, G. A., Smith, R. B.: Transitional cell carcinoma of the bladder associated with cyclophosphamide. J. Urol. *112*, 603 (1974)

DeVita, V. T., Arseneau, J. C., Sherins, R. J., Canellos, G. P., Young, R. C.: Intensive chemotherapy for Hodgkin's disease: Long-term complications. Nat. Cancer Inst. Monogr. *36*, 447 (1973)

Evans, S., Waters, L., Hardin, B., Matossian, N., Goltra, E., de Luca, V.: Carcinoma primitif du foie, apparu chez un sujet atteint de leucémie myéloide traitée par uréthane. Rev. Hémat. *6*, 148 (1951)

Frick, E., Angstwuren, H., Strauss, G.: Immunosuppressive Therapie der multiplen Sklerose. Muench. med. Wschr. *116*, 1987 (1974)

Gärtner, U., Schief, A., Stocker, W. G.: Beitrag zum Problem maligner Zweiterkrankungen nach zytostatischer Therapie. Muench. med. Wschr. *117*, 671 (1975)

Gutjahr, P., Spranger, J.: Acute leukemia following anticancer treatment. J. Pediat. *87*, 1004 (1975)

Hartwich, G.: Nebenwirkungen zytostatischer und immunosuppressiver Therapie. Fortschr. Med. *91*, 357 (1973)

Hartwich, G., Butzler, W.: Zweittumoren bei Hämoblastosen unter zytostatischer Therapie. Med. Klin. *66*, 1445 (1971)

Heck, J.: Endoxanbehandlung der primär chronischen Polyarthritis. Med. Welt *22*, 1077 (1971)

Heckner, F.: Klinische Demonstrationen. Klin. Wschr. *37*, 1051 (1959)

Hueper, W. C., Wiley, F. H., Wolte, H. D.: Experimental production of bladder tumours in dogs by administration of beta-naphthylamine. J. industr. Hyg. *20*, 46 (1948)

Hunstein, W., von Rehn, K.: Tumorinduktion durch Zytostatika beim Menschen. Dtsch. med. Wschr. *100*, 155 (1975)

IARC Monograph Series on the Evaluation of Carcinogenic Risk in Man. Vol. IX. Lyon: International Agency for Research on Cancer, 1975

---

[3] For further reviews on iatrogenic carcinogenesis by cytostatic agents see Thiede (1964), Aron (1968), Heck (1971), O'Gara (1971), Hartwich (1973), Frick et al. (1974), Vormittag (1974), Schmähl (1974, 1975)

Kahlau, G.: Der Lungenkrebs. Ergebn. Path. path. Anat. 37, 258 (1954)

Karchner, R. K., Amare, M., Larsen, W. E., Mallouk, A. G., Caldwell, G. G.: Alkylating agents as leukemogens in multiple myeloma. Cancer (Philad.) 33, 1103 (1974)

Klemm, D., Merker, H., Westerhausen, M.: Akute Neuproliferation bei paraproteinämischen Hämoblastosen unter langdauernder Zytostatikabehandlung. Dtsch. med. Wschr. 93, 2063 (1968)

Kyle, R. A., Pierre, R. V., Bayrd, E. D.: Multiple myeloma and acute myelomonocytic leukemia. Report of four cases possibly related to melphalan. New Engl. J. Med. 283, 1121 (1970)

Labedski, L., Fuchs, H. J., Grips, K. H.: Plasmocytom und finale unreifzellige Leukose. Dtsch. med. Wschr. 101, 108 (1976)

Maltoni, C., Chieco, P.: Adriamicina: Un nuovo potente cancerogeno. Gli Ospedali della Vita 2, 107 (1975)

Michot, E.: Beitrag zum Problem des Plasmocytoms ohne Paraproteine. Dtsch. med. Wschr. 93, 2072 (1968)

Min, K. W., Györkey, F.: Interstitial pulmonary fibrosis, atypical epithelial changes, and bronchiolar cell carcinoma following busulfan therapy. Cancer (Philad.) 22, 1027 (1968)

Nelson, B.M., Andrews, G. A.: Breast cancer and cytologic dysplasia in many organs after busulfan (Myleran). Amer. J. clin. Path. 42, 37 (1964)

O'Gara, R. W., Adamson, R. H., Kell, M. G., Dalgard, D. W.: Neoplasms on the hematopoietic system in nonhuman primates. Report of tumors and two leukemias induced by procarbazine. J. nat. Cancer Inst. 46, 1121 (1971)

Philips, F. S., Gilladoga, A., Marquardt, H., Sternberg, S. S., Vidal, P. M.: Some observations on the toxicity of adriamycin. Cancer Chemother. Rep. 6, 177 (1975)

Rehn, L.: Blasengeschwülste bei Fuchsin-Arbeitern. Arch. klin. Chir. 50, 588 (1895)

Reis, H. E., Hossfeld, D. K., Stier, H. W.: Über die Kombination von malignem Tumor und Hämoblastose. Neoplastische Transformation nach Therapie oder Zweiterkrankung. Med. Welt 2, 2411 (1969)

Rieche, K.: Zur Frage der Kanzerogenität von Zytostatika. Arch. Geschwulstforsch. 34, 240 (1969)

Rupprecht, L., Blessing, M. H.: Fibrosarkom der Harnblase nach siebenjähriger Chemotherapie einer Lymphogranulomatose im Kindesalter. Dtsch. med. Wschr. 98, 1663 (1973)

Scherf, H. R., Schmähl, D.: Experimental investigations on immunosuppressive properties of carcinogenic substances in male Sprague-Dawley-rats. Recent Results Cancer Res. 52, 76 (1975)

Schmähl, D.: Karzinogene Wirkung von Cyclophosphamid und Triazichon bei Ratten. Dtsch. med. Wschr. 92, 1150 (1967)

Schmähl, D.: Entstehung, Wachstum und Chemotherapie maligner Tumoren. Aulendorf: Editio Cantor 1970, 2 nd ed.

Schmähl, D.: Karzinogenese und Immunodepression. Dtsch. med. Wschr. 45, 1771 (1971)

Schmähl, D.: Investigations on the influence of immunodepressive means on the chemical carcinogenesis in rats. Z. Krebsforsch. 81, 211 (1974)

Schmähl, D.: Experimental investigations with anticancer drugs for carcinogenicity with special reference to immunodepression. Recent Results Cancer Res. 52, 18 (1975)

Schmähl, D., Osswald, H.: Experimentelle Untersuchungen über carcinogene Wirkungen von Krebs-Chemotherapeutika und Immunosuppressiva. Arzneimittel-Forsch. 20, 1461 (1970)

Schmähl, D., Mundt, D., Schmidt, K. G.: Experimental investigations on the influence upon the chemical carcinogenesis. 1st communication: Studies with ethylnitrosourea. Z. Krebsforsch. *82,* 91 (1974)

Schmähl, D., Habs, M.: Absence of carcinogenic effects of estradiol mustard (NSC-112259) in rats. Z. Krebsforsch. *87,* 197 (1976)

Schmalzl, F., Keiser, G., Kresbach, E., Asamer, H., Braunsteiner, H.: Plasmozytom, Alkylantien und akute myelogene Leukämien. Dtsch. med. Wschr. *100,* 1961 (1975)

Scott, W. W., Boyd, H. L.: A study of the carcinogenic effect of betanaphthylamine on the normal and substituted sigmoid loop ladder of dogs. J. Urol. *70,* 914 (1953)

Shimkin, M. B., Weisburger, J. H., Weisburger, E. K., Gubareff, N., Suntzeff, V.: Bioassay of 29 alkylating chemicals by the pulmonary-tumor response in strain A mice. J. Nat. Cancer Inst. *36,* 915 (1966)

Sieber, S. M., Adamson, R. H.: Toxicity of antineoplastic agents in man. Chromosomal aberrations, antifertility effects, congenital malformations, and carcinogenic potential. Advanc. Cancer Res. *22,* 57–144 (1975)

Sotrel, G., Jafari, K., Lash, A. F., Stepto, F., Stepto, R. C.: Acute Leukemia in advanced ovarian carcinoma after treatment with alkylating agents. Obstet. Gynec. *47,* 67 (1976)

Stavem, P., Harboe, M.: Acute erythroleukaemia in a patient with malphalan for the cold agglutinin syndrome. Scand. J. Haemat. *8,* 375 (1971)

Storti, E., Mauri, C., Artusi, T., Traldi, A., Vaccari, G. L.: Zusammentreffen von Leukämie und neoplastischen Lymphopatien. Muench. med. Wschr. *109,* 1597 (1967)

Terbrüggen, A.: Neoplastische Retikulose nach zytostatischer Dauerbehandlung von radikal operierten Karzinomen. Verh. Dtsch. Ges. Path. *49,* 241 (1965)

Thiede, T., Chievitz, E., Christensen, B. Chr.: Chlornaphazine as a bladder carcinogen. Acta med. Scand. *175,* 721 (1964)

Thiede, T., Christensen, B. Chr.: Bladder tumours induced by chlornaphazine. Acta med. scand. *185,* 133 (1969)

Videbaek, A.: Chlornaphazin (Erysan^R) may induce cancer of the urinary bladder. Acta med. scand. *176,* 45 (1964)

Vormittag, W.: Zytostatische Therapie, chromosomale Aberrationen und karzinogene Wirkung. Wien. klin. Wschr. *86,* 69 (1974)

Wada, S., Miyanishi, M., Nishimoto, Y., Kambe, S.: Mustard gas as a cause of cancer in man. Lancet, *I, 68,* 1161 (1968)

Wall, R. L., Clausen, K. P.: Carcinoma of the bladder in patients receiving cyclophosphamide. New Engl. J. Med. *293,* 271 (1975)

Weisburger, J. H., Griswold, D. P., Prejeau, J. D., Casey, A. E., Wood, H. B., Weisburger, E. K.: The carcinogenic properties of some of the principal drugs used in clinical cancer chemotherapy. Recent Results Cancer Res. *52,* 1 (1975)

Weiss, A., Weiss, B.: Karzinogenese durch Lost-Exposition beim Menschen, ein

Wildhack, R.: Monozytenleukämie mit G-Paraproteinämie nach Röntgenbestrahlung und zytostatischer Behandlung eines Myosarkoms. Dtsch. med. Wschr. *92,* 255 (1967)

Worth, P. H. L.: Cyclophosphamide and the bladder. Brit. med. J. *3,* 182 (1971)

Organ Transplantation

It was observed that the cancer incidence in recipients of organ transplants was uncommonly increased. In several cases, in which the donor died with cancer, the tumor in the recipient originated from the transplanted organ (Martin et al., 1965; for further references see Penn, 1970 and 1975). The histologic features of both tumors were therefore identical. It was proposed that in the transplanted organ, which appeared normal at the time of transplantation, "dormant" cancer cells were present that started multiplying, possibly because of the massive immunodepressive treatment that was necessary to prevent the rejection. Here we have the first historic case of a transplantation tumor in man. The formation of these tumors is not connected with the original process of carcinogenesis, i.e. the change of normal cells into cancer cells, but covers a process of growth and multiplication of already present, transplanted cancer cells into a clinically manifest tumor.

This type of tumor formation is not unprobable if one considers the fact that experimental studies had shown that in animals pretreated with irradiation or immunodepressive cytostatics, the growth of transplantable tumors was enhanced and the incidence of metastases increased (Schmähl and Stutz, 1962; Schmähl, 1963; Schmähl and Sattler, 1964; for further references see Schmähl, 1970).

Organ transplants from healthy donors also increased the cancer incidence. About 5% of the organ transplant recipients developed malignant tumors (predominantly lymphomas, but also solid carcinomas) 6 months to 3 years after the transplantation (McKhann, 1969; Penn and Starzl, 1970; Allison, 1970; Gatti and Good, 1971; Enderlin and Guysan, 1973; Penn, 1975). Compared to the normal, age-adjusted population, the cancer incidence in organ transplant recipients is increased 50–100-fold.

We have so far no explanation for this phenomenon. It is possible that the immunodepressive treatment disturbes the immunologic surveillance mechanisms of the organism and initiates the growth of dormant tumor cells that may occur physiologically in the body. It is also possible that oncogenic viruses, which normally are controlled by the immune system, increase after immunodepressive treatment and thus lead to the formation of tumor. This is supported by the finding that predominantly lymphomas were formed. Another theory is chemical carcinogenesis. Several immunodepressive agents used in organ transplantation have to be considered as chemical carcinogens. The strikingly short latent periods of the tumors (19 months, on the average), however, contradict this view. Chemically induced tumors commonly have latency periods of several years or even decades.

The supposition that immunologic mechanisms play a role is supported by the finding that inborn immune deficiency is frequently associated with an increased incidence of malignant tumors (Page et al., 1963; Haerer et al., 1969; Levin and Perlov, 1971; Waldmann, 1972; Hamoudi et al., 1974). In children suffering from Wiscott-Aldrich syndrome, agammaglobulinemia, or ataxia teleangiectasia, the incidence of malignant tumors was increased up to 10,000-fold the normal value.

Experimental studies on the influence of the immune system on carcinogenesis showed contradictory results. Whereas the growth of viral tumors was promoted by immunodepressive agents, chemically-induced tumors responded only to a less extent or not at all to the immunodepressive treatment (reviews by Gleichmann and Gleichmann, 1973; Andrews, 1974; Kroes et al., 1975). In our own investigations the initiation and the growth of chemically induced tumors was not influenced by immunodepressive agents (Schmähl, 1971, 1974, 1975; Schmähl, et al., 1974, 1976a, b; Scherf and Schmähl, 1975; Scherf et al., 1970).

Since the time of Paul Ehrlich it has been speculated again and again that the immune mechanisms might influence the growth and also possibly the formation of tumors. The observations in organ transplant recipients and patients with inborn immune deficiency indicationed such an association. These observations stimulated a series of studies that possibly will contribute to our understanding of the role of immune processes in carcinogenesis.

References

Allison, A. C.: Tumor development following immunosuppression. Proc. Roy. Soc. Med. 63, 1077 (1970)

Andrews, E. J.: Failure of immunosurveillance against chemically induced in situ tumors in mice. J. Nat. Cancer Inst. 52, 729 (1974)

Enderlin, F., Guysan, Y.: Maligne Tumoren unter Immunosuppressiva, ein neues Problem beim Transplantationsempfänger. Helv. Chir. Acta 40, 773 (1973)

Gatti, R. A., Good, R. A.: Occurrence of malignancy in immunodeficiency diseases. Cancer (Philad.) 28, 89-97 (1971)

Gleichmann, A., Gleichmann, E.: Immunosuppression and neoplasia. Klin. Wschr. 51, 255 (1973)

Haerer, A. F., Jackson, J. F., Evers, C. G.: Ataxia teleangiectasia with gastric adeno-carcinoma. J. Amer. Med. Ass. 210, 1884 (1969)

Hamoudi, A. B., Ertel, J., Newton, W. A., Reiner, C. B., Clatworthy, H. W.: Multiple neoplasm in a child associated with IGA deficiency. Cancer (Philad.) 33, 1134 (1974)

Kroes, R., Weiss, J. W., Weisburger, J. H.: Immune suppression and chemical carcinogenesis. Recent Results Cancer Res. 52, 65 (1975)

Levin, S., Perlov, S.: Ataxia-Teleangiectasia. Israel J. Med. Sci. *7*, 1535 (1971)

Martin, D. C., Rubin, M., Rosen, V. J.: Cadaveric renal homotransplantation with inadvertent transplantation of carcinoma. J. Amer. Med. Ass. *192*, 82 (1965)

McKhann, C. F.: Primary malignancy in patients undergoing immunosuppression for renal transplantation. Transplantation *8*, 209 (1969)

Page, A. R., Hansen, A. E., Good, R. A.: Occurrence of leukemia and lymphoma in patients with agammaglobulinemia. Blood *21*, 197 (1963)

Penn, I.: Malignant tumors in organ transplant recipients. Recent Results Cancer Res. *35*, 1 (1970)

Penn, I.: Malignant diesease in immunodeficient states in man. Recent Results Cancer Res. *52*, 96 (1975)

Penn, I., Starzl, T. E.: Malignant lymphomas in transplantation patients. Int. J. Clin. Pharmacol. *3*, 49 (1970)

Scherf, H. R., Krüger, C., Karsten, C.: Untersuchungen an Ratten über immunosuppressive Eigenschaften von Cytostatica unter besonderer Berücksichtigung der carcinogenen Wirkung. Arzneimittel-Forsch. *20*, 1467 (1970)

Scherf, H. R., Schmähl, D.: Experimental investigations on immunodepressive properties of carcinogenic substances in male Sprague-Dawley-rats. Recent Results Cancer Res. *52*, 76 (1975)

Schmähl, D.: Wert und Gefahr der Krebschemotherapie. Dtsch. med. Wschr. *88*, 1463 (1963)

Schmähl, D.: Entstehung, Wachstum und Chemotherapie maligner Geschwülste. Aulendorf: Editio Cantor, 1970, 2nd ed.

Schmähl, D.: Karzinogenese und Immunodepression. Dtsch. med. Wschr. *45*, 1771 (1971)

Schmähl, D.: Investigations on the influence of immunodepressive means on the chemical carcinogenesis in rats. Z. Krebsforsch. *81*, 211 (1974)

Schmähl, D.: Experimental investigations with anticancer drugs for carcinogenicity with special reference to immunodepression. Recent Results Cancer Res. *52*, 18 (1975)

Schmähl, D., Danisman, A., Habs, M., Diehl, B.: Experimental investigations upon the chemical carcinogenesis. 3rd communication: Studies with 1,2-dimethylhydrazine. Z. Krebsforsch. *86*, 89–94 (1976)

Schmähl, D., Habs, M., Diehl, B.: Experimental investigations on the influence upon the chemical carcinogenesis. 2nd. communication: Studies with 3,4-benzopyrene. Z. Krebsforsch. *86*, 85–88 (1976)

Schmähl, D., Mundt, D., Schmidt, K. G.: Experimental investigations on the influence upon chemical carcinogenesis. 1st communication: Studies with ethylnitrosourea. Z. Krebsforsch. *82*, 91 (1974)

Schmähl, D., Sattler, W.: Der Einfluß der Vorbehandlung von Ratten mit alkylierenden Substanzen und anderen Giften auf das Wachstum des Yoshida-Sarkoms. Arzneimittel-Forsch. *14*, 746 (1964)

Schmähl, D., Stutz, E.: Abhängigkeit der Tumorentwicklung bei Ratten nach i.v. Injektion des Yoshida-Aszites-Sarkoms von vorausgegangener Röntgenganzkörperbestrahlung. Naturwissenschaften *49*, 424 (1962)

Waldmann, T. A.: Immunodeficiency disease and neoplasia. Ann. Intern. Med. *77*, 605 (1972)

Iron-Dextran Complex

Experimental induction of fibrosarcomas by injection of iron-dextran complexes (Richmond, 1959; Golberg, 1960; Haddow and Horning,

1960) posed the question of whether man also might be endangered by this medication.

In a survey of the literature we found five cases of tumors possibly associated with the use of iron-dextran complexes, which are described below.

Crowley and Still (1960) described an anemic woman with carcinoma of the cervix, who had been injected intramuscularly with Imferon (total dose: 2.5 g). Eight weeks after the start of treatment a metastasis was detected at the injection site. The authors concluded that this metastasis had been furthered by a reactive Imferon-induced hyperplasia of the subcutaneous fat tissue.

A similar case, indicating the formation of metastasis in traumatically impaired tissue, was described by Worthy and Wynne (1960), who observed a metastatic carcinoma at the application site of a penicillin-oil-suspension. According to Willis (1952) the formation of metastasis in traumatically impaired tissue is theoretically possible, though rare.

Robinson et al. (1960) described a 74-year-old woman who had been injected with 600 mg iron dextran in the deltoid muscle. A local sarcoma of the soft parts developed 3 years after the end of treatment. The authors did not exclude, however, a metastasis of an occult sarcoma.

MacKinnon and Bancewicz (1973) observed a reticulum cell sarcoma in a 62-year-old woman, following a series of injections of an "iron-sorbital-citric acid complex." In a second patient they found pleomorphic sarcomas occurring 2 and 5 years after injection of 2 g iron dextran.

The experimental doses of iron-dextran complexes that led to local fibrosarcomas in rats and mice were more than 100-fold greater than those given to humans (Thedering, 1964). From the experimental findings it can be concluded that the local carcinogenicity of iron dextran is less dose-related than dependent on dose/body weight ratio, animal species, life expectancy of animals, and mode of application (Haddow and Horning, 1960; Thedering, 1964; MacKinnon and Bancewicz, 1973).

In England iron dextran was initially withdrawn from the market, but was reinstituted after a further thorough investigation. It can hardly be assumed that low therapeutic doses of iron dextran (1–3 g) have carcinogenic effects. Thedering (1964) emphasized that no local sarcomas were found during 10 years' observation of 3 million iron dextran-treated patients. The isolated cases described in the literature might therefore be coincidental (MacKinnon and Bancewicz, 1973).

References

Crowley, J. D., Still, W. J. S.: Metastatic carcinoma at the site of injection of iron-dextran complex. Brit. Med. J. *1*, 1411 (1960)

Golberg, L.: cited by Thedering, F. (1964)

Haddow, A., Horning, E. S.: cited by Thedering, F. (1964)

MacKinnon, A. E., Bancewicz, J.: Sarcoma after injection of intramuscular iron. Brit. Med. J. *2*, 277 (1973)

Richmond, H. G.: Induction of sarcoma in the rat by iron-dextran complex. Brit. med. J. *2*, 277 (1973)

Robinson, C. E. G., Bell, D. N., Sturdy, J. H.: Possible association of malignant neoplasm with iron-dextran injection. Brit. med. J. *2*, 648 (1960)

Thedering, F.: Die Verträglichkeit der Eisentherapie unter Berücksichtigung der karzinogenen Wirkung des Dextraneisenkomplexes. Med. Welt *15*, 277 (1964)

Willis, R. S.: The spread of tumours in the human body. London: Butterworth, 1952

Worthy, T. S., Wynne, E. J. C.: Metastastic carcinoma at the site of injection of penicillin. Brit. Med. J. *2*, 1208 (1960)

Hydantoin Derivatives

Hydantoin derivatives are used mainly for the treatment of epilepsy. They have a series of undesired side-effects, such as gastrointestinal disorders, toxicity to the central nervous system, and gingival hyperplasia.

Saltzstein and Ackermann (1959) observed malignant lymphomas following this medication, and pseudolymphomas histologically mimicking lymphosarcomas or Hodgkin's disease. They found lymphadenopathies in seven patients who had been treated with diphenylhydantoin; one of them, who later died with generalized lymphoma, had received the drug over a period of 7 years. According to Leder and Lennert (1972), however, this case can hardly be attributed to diphenylhydantoin because no histologic correspondence was observed between the lymphadenopathy and a sarcoma that developed 4 years later.

Hyman and Sommers (1966) observed three Hodgkin's lymphomas and three malignant lymphomas that occurred during diphenylhydantoin treatment. The drug had been applied over a period of 2 to 17 years. Two patients died with lymphosarcomas. The others showed improvement of their disease after radiation or cytostatic treatment and, with one exception, diphenylhydantoin treatment was continued.

Gams et al. (1968) described one lymphadenopathy, histologically corresponding to malignant lymphoma, that occurred during diphenylhydantoin treatment. The lymphadenopathy receded after diphenylhydantoin was discontinued. One year later the patient died with malignant lymphoma.

Alberto et al. (1970) observed one benign lymphoma and two Hodgkin's lymphomas after 5 to 10 years' treatment with diphenylhydantoin.

Rausing and Trell (1971) described a woman who died with malignant reticulosis following 3 years' treatment with diphenylhydantoin. Histologic examination revealed malignant lymphogranulomatosis and reticulum cell hyperplasia involving the lymph nodes, liver, spleen, and bone marrow.

Wildhack (1973) observed three stem cell leukoses following diphenylhydantoin treatment.

Charlton and Lunsford (1971) demonstrated phenytoin anamnesis, ranging from 2 to 21 years, in 7 out of 300 patients with malignant lymphoma; in the control population they found only one certain and one questionable lymphoma. Histologic examination of the malignant lymphomas showed 2 Hodgkin's lymphomas and 5 reticulum cell sarcomas.

A review of hydantoin-related lymphadenopathy, including a discussion of malignant lymphomas, is given by Beil and Prechtel (1973)[4].

Charlton and Lunsford (1971) demonstrated phenytoin use in 2.6% of their patients. Assuming that in the United States all epileptics, who account for 0.4% of the population, were being treated with this drug, we would have a control group of 0.3–0.6% diphenylhydantoin-treated patients. This difference is statistically significant. On the other hand, Clemmesen et al. (1974) found no increased incidence of lymphomas in patients treated with hydantoin derivatives or barbiturates.

It is believed that phenytoin acts on the immune system. Krüger (1970), on the basis of his experimental findings, supposed that phenytoin had no direct carcinogenic activity, but thought the chronic antigen stimulus and decrease of immunologic reactivity might be responsible for the formation of malignant lymphomas. Krüger et al. (1972) induced generalized lymphosarcomatoses by dilantin in mice, and Juhasz et al. (1970) induced lymphosarcomas, reticulosarcomas, and leukoses by long-term treatment with diphenylhydantoin in 10 out of 40 rats. These experimental findings are interesting in that five human leukemias were observed after diphenylhydantoin treatment (Stich and Ehrhart, 1963; Gebert and Lorenz, 1968; Wildhack, 1973).

The evaluation of hydantoin derivatives with respect to possible carcinogenic effects in man is very difficult. Such effects, however,

---

[4] (See also Leading articles (Lancet, 1971) and McCarthy and Chalmers, 1964)

cannot be excluded. We therefore suggest that the range of indications of these substances, which have been used rather liberally, be more confined.

References

Alberto, P., Cougn, R., Maurice, P., Weber, J.: Trois cas de lymphome malin ou pseudolymphome chez des épileptiques traités par la diphénylhydantoine. Schweiz. med. Wschr. *101*, 1773 (1970)

Beil, E., Prechtel, K.: Malignes Lymphom oder Hydantoin-Lymphadenopathie. Muench. med. Wschr. *115*, 2033 (1973)

Charlton, M. H., Lunsford, E.: Cited in Leading Articles. Lancet, 1971

Clemmesen, J., Fuglsang-Frederiksen, V., v. Pleim, C. M.: Are anticonvulsants carcinogenic? Lancet 1974, 705–707

Gams, R. A., Neal, J. A., Conrad, F. G.: Hydantoin-induced pseudolymphoma. Ann. Intern. Med. *69*, 557 (1968)

Gebert, P., Lorenz, K.: Maligne Hämoblastosen nach prophylaktischen und therapeutischen Maßnahmen. Dtsch. Gesundh.-Wes. *23*, 1318 (1968)

Hyman, G. A., Sommers, S. C.: The development of Hodgkin's disease and lymphoma during anticonvulsant therapy. Blood *28*, 416 (1966)

Juhasz, J., Balo, J., Szende, B.: Tumors induced by diphenylhydantoin. Acta Morph. Acad. Sci. Hung. *18*, 147 (1970)

Krüger, G.: Zur Pathogenese von Tumoren des lymphoretikulären Gewebes bei Transplantationsempfängern. Verh. Dtsch. Ges. Path. *54*, 175 (1970)

Krüger, G., Harris, D., Sussman, E.: Effect of dilantin in mice. II. Lymphoreticular tissue atypia and neoplasia after chronic exposure. Z. Krebsforsch. *78*, 290 (1972)

Leading Articles: Is phenytoin carcinogenic? Lancet (1971) 1071

Leder, L. D., Lennert, K.: Über iatrogene Lymphknotenveränderungen. Verh. Dtsch. Ges. Path. *54*, 310 (1972)

McCarthy, D. D., Chalmers, T. M.: Haematological complications of phenylbutazone therapy. Review of the literature and report of two cases. Canad. Med. Ass. J. *90*, 1061 (1964)

Rausing, A., Trell, E.: Malignant lymphogranulomatosis and anticonvulsant therapy. Acta Med. Scand. *189*, 131 (1971)

Saltzstein, S. L., Ackermann, L. V.: Lymphadenopathia induced by anticonvulsant drugs mimicking clinically and pathologically malignant lymphomas. Cancer (Philad.) *12*, 164 (1959)

Stich, W., Ehrhart, H. (1963): cited by Wildhack, R. (1973)

Wildhack, R.: Leukämie und Hydantoin-Behandlung. Muench. med. Wschr. *115*, 1275 (1973)

Phenacetin

Phenacetin is contained in most analgesics. It is well known that excessive use of phenacetin causes interstitial nephritis and papillary necrosis (Spühler and Zollinger, 1953; Hultengren, 1961).

In 1965, Hultengren et al. drew attention to a possible association between phenacetin overconsumption and carcinoma of the renal pelvis.

Bengtsson et al. (1968) found 8 renal pelvis carcinomas and 2 carcinomas of the urinary bladder among 104 patients with chronic nonobstructive nephritis following analgesic abuse. They proved phenacetin use in 5 out of 28 patients with renal pelvis carcinoma.

Angervall et al. (1969), during the years 1960 to 1968, observed 15 patients with carcinoma of the renal pelvis. Ten, possibly 12, of them had consumed excessive amounts of analgesics. Nine patients had been employees at an arms factory in which heavy analgesic consumption was common.

Hobye and Nielsen (1971) found 2 carcinomas of the renal pelvis among 101 phenacetin users.

Bock and Hogrefe (1972) found one case of phenacetin overconsumption among 31 patients with renal pelvis carcinoma. Renal pelvis carcinomas, according to these investigators, are thus far rare in Germany. However, since analgesic consumption has been increasing since the end of the war, and assuming a latent period of about 20 years, we would expect an increase in incidence during the next few years.

Leistenschneider and Ehmann (1973), in the years 1960–1969, when analgesic abuse in Switzerland was at its height, found 17 renal pelvis carcinomas among 21, 291 autopsies. Phenacetin use was proved in eight cases. During the initial phase of analgesic abuse, between 1950 and 1959, they found 11 renal pelvis carcinomas among 15,635 autopsies. Only one questionable carcinoma of the renal pelvis was observed among 9225 autopsies carried out from 1925 to 1934. The differences found in the individual phases of analgesic abuse are statistically significant.

In a survey of the literature we found 38 cases with tumors of the uriary tract following excessive use of phenacetin; these are shown in Table 10. Thirty-two tumors were formed in the renal pelvis, two were situated in the urinary bladder, and two patients had tumors of both organs. Histologically the tumors were carcinomas and papillomas of the renal epithelium (26 and ten cases, respectively), and carcinomas and papillomas of the epithelium of the urinary bladder (four and two cases, respectively). Papillary necrosis and chronic interstitial nephritis were found in 31 and 32 patients, respectively. The average latent period of the tumors was 20 years. Phenacetin ingestion ranged from 1–25 kg per patient.

The average age of patients was 57 years as compared to 63 (Newman, 1967) and 65 years (Grace, 1967) for renal pelvis carcinomas of other origins. The male: female ratio was 1:1, whereas it is normally 3.5:1 (Deming, 1963; Riches, 1967). This difference is attributed to the fact that analgesic abuse is more common in women.

Chemically, phenacetin is p-ethoxyacetanilide. The largest part of an ingested dose is dealkylated to N-acetyl-p-aminophenol (Goodman and Gilman, 1970), and only a small fraction is deacetylated to phenetidin, which, by esterification with sulfate, is transformed to o-amino-phenol (p-ethoxy-o-sulfonyl-oxy-aniline) (Büch et al., 1966).

It is not known whether phenacetin or one of its metabolic products has direct carcinogenic activity, or whether tumor formation is only indirectly favored by a sharp restimulation in the course of phenacetin-induced chronic nephritis.

Bengtsson and Angervall (1970) suggested that phenacetin might be responsible because 2-hydroxyphenetidin, after transferance to o-amino-phenol, has a chemical structure similar to some known bladder carcinogens in man. Excretion of this metabolite depends significantly on the ingested dose and is increased to ten times the normal value when high doses are applied over a short period (Dubach and Raaflaub, 1969). The main metabolite of phenacetin, N-acetyl-p-aminophenol, is excreted as glucuronid to the extent of 28% and as sulfate to the extent of 15% (Büch et al., 1966). During short-term application of high phenacetin doses and simultaneous increase of β-glucuronidase activity accompanying urothelial injuries (Fripp, 1965; Leistenschneider and Ehmann, 1973), large amounts of decoupled acetyl-p-aminophenol could become free (Dubach and Raaflaub, 1969). One could speculate that this metabolite is chemically related to 4-amino-diphenyl, which is a known bladder carcinogen in man.

It is also possible that chronic inflammation of the epithelium and extremely increased regeneration, accompanying chronic nephritis and papillary necrosis, favor tumor formation (Hultengren et al., 1965; Leistenschneider and Ehmann, 1973).

We have not yet been able to induce cancer experimentally with phenacetin (Schmähl and Reiter, 1954). This indicates that the drug is metabolized differently by man and animals (Schmähl, 1972). It is also interesting that, despite massive phenacetin use, only a small percentage of human individuals react with renal lesions (Landmann-Suter and Dubach, 1975; Landmann et al., 1975).

References

Adam, W. R., Dawborn, J. K., Price, C. G., Riddel, J., Story, H.: Anaplastic transitional-cell carcinoma of the renal pelvis in association with analgesic abuse. Med. J. Aust. *57*, 1108 (1970)

Angervall, L., Bengtsson, U., Zetterlund, G., Zsigmond, M.: Renal pelvic carcinoma in a Swedish district with abuse of a phenacetin-containing drug. Brit. J. Urol. *41*, 401 (1969)

Begley, M., Chadwick, J. M., Jepson, R. P.: A possible case of analgesic abuse associated with transitional-cell carcinoma of the bladder. Med. J. Aust. *57*, 1133 (1970)

Bengtsson, U., Angervall, L.: Analgesic abuse and tumours of the renal pelvis. Lancet *1*, 305 (1970)

Bengtsson, U., Angervall, L., Ekmann, H., Lehmann, L.: Transitional-cell tumours of the renal pelvis in analgesic abusers. Scand. J. Urol. Nephrol. *2*, 145 (1968)

Bock, K. D., Hogrefe, J.: Analgeticaabusus und maligne Tumoren der ableitenden Harnwege. Muench. med. Wschr. *114*, 645 (1972)

Büch, H., Häuser, H., Pfleger, K., Rüdiger, W.: Über die Ausscheidung eines noch nicht beschriebenen Phenacetinmetaboliten beim Menschen und bei der Ratte. Naunyn-Schmiedebergs Arch. Exp. Path. Pharmak. *253*, 25 (1966)

Deming, C. L.: Tumors of the kidney. In: Urology. Campbell, M. F. (ed.) London – Philadelphia: W. B. Saunders Co. 1963

Dubach, U. C., Raaflaub, J. (1969): cited by Leistenschneider, W., Ehmann, R. (1973)

Fripp, D. J.: The origin of urinary-beta glucuronidase. Brit. J. Cancer *19*, 330 (1965)

Goodman, L. S., Gilman, A.: The pharmacological basis of therapeutics. New York: The Macmillan Co. 1970

Grace, D. A.: Carcinoma of the renal pelvis: A 15 year review. J. Urol. *98*, 566 (1967)

Grob, H. U.: Phenacetinabusus und Nierenbeckencarcinom. Helv. Chir. Acta *38*, 537 (1971)

Hobye, G.,ielsen, O. E.: Renal pelvic carcinoma in phenacetin abusers. Scand. J. Urol. Nephrol. *5*, 190 (1971)

Hultengren, N.: Renal papillary necrosis, a clinical study of 103 cases. Acta Chir. Scand. Suppl. 277 (1961)

Hultengren, N., Lagergren, C., Ljungvist, A.: Carcinoma of the renal pelvis in renal papillary necrosis. Acta Chir. Scand. *130*, 314 (1965)

Landmann, K. G., Rutishauser, G., Dubach, U. C.: Phenacetin-Abusus und Harnwegs-tumoren. Urologe A *14*, 75 (1975)

Landmann-Suter, R., Dubach, U. C.: Phenacetin und Nierenschädigung. Dtsch. med. Wschr. *100*, 1451 (1975)

Leistenschneider, W., Ehmann, R.: Nierenbeckencarcinom nach Phenacetinabusus. Schweiz. med. Wschr. *103*, 433 (1973)

Mannion, R. A., Susmano, D.: Phenacetin abuse causing bladder tumour. J. Urol. *106*, 602 (1971)

Newman, D. M.: Transitional cell carcinoma of the upper urinary tract. J. Urol. *98*, 322 (1967)

Riches, E.: Tumors of the kidney and ureter. In: Handbuch der Urologie. Alken, C. E., Dix, V. W., Weyrauch, H. M., Wildholz, E. (eds.): Berlin – Heidelberg – New York: Springer 1967, Vol. XI/1,1

Schmähl, D.: Toxikologische Probleme der iatrogenen Carcinogenese. Verh. Dtsch. Ges. Path. *56*, 133 (1972)

Schmähl, D., Reiter, A.: Fehlen einer cancerogenen Wirkung beim Phenacetin. Arzneimittel-Forsch. *4*, 404 (1954)

Spühler, O., Zollinger, H. U.: Die chronische interstitielle Nephritis. Z. klin. Med. *151*, 1 (1953)

Table 10. Tumors of the renal pelvis associated with phenacetin overconsumption

Patient Age	Sex	Type of tumor	Reference
52	F	Carcinoma of the renal epithelium	Hultengren et al., 1965
60	F	Carcinoma of the renal epithelium and papilloma of the urinary bladder	,,
51	F	Carcinoma of the renal epithelium	,,
58	F	Carcinoma of the renal epithelium and papilloma of the urinary bladder	,,
42	F	Papillomatous carcinoma of the renal epithelium	,,
44	M	Carcinoma of the renal epithelium	Bengtsson et al., 1968
44	F	Carcinoma of the renal epithelium	,,
44	F	Carcinoma of the renal epithelium	,,
52	F	Carcinoma of the renal epithelium	,,
55	F	Carcinoma of the renal epithelium	,,
56	M	Carcinoma of the renal epithelium	,,
56	M	Papilloma of the renal epithelium	,,
58	F	Carcinoma of the renal epithelium	,,
58	F	Carcinoma of the renal epithelium	,,
59	F	Renal papilloma	,,
64	M	Papilloma of the renal epithelium	,,
66	F	Papilloma of the renal epithelium	,,
68	F	Papilloma of the renal epithelium	,,
44	M	Carcinoma of the renal epithelium	Angervall et al., 1969
44	M	Papilloma of the renal epithelium	,,
55	M	Carcinoma of the renal epithelium	,,
57	M	Papilloma of the renal epithelium	,,
58	M	Carcinoma of the renal epithelium	,,
62	M	Papilloma of the renal epithelium	,,
63	M	Carcinoma of the renal epithelium	,,
64	M	Papilloma of the renal epithelium and carcinoma of the urinary bladder	,,
66	M	Papilloma of the renal epithelium and carcinoma of the urinary bladder	,,
67	M	Carcinoma of the renal epithelium	,,
70	M	Carcinoma of the renal epithelium	,,

Table 10 (continued)

Patient Age	Sex	Type of tumor	Reference
78	M	Carcinoma of the renal epithelium	Angervall et al., 1969
40	F	Epithelial carcinoma of the urinary bladder	Begley et al., 1970
37	F	Carcinoma of the renal epithelium	Adam et al., 1970
50	F	Carcinoma of the renal epithelium	Grob, 1971
50	F	Carcinoma of the renal epithelium	,,
67	M	Carcinoma of the renal epithelium	Hobye and Nielsen, 1971
65	F	Carcinoma of the renal epithelium	,,
48	M	Epithelial carcinoma of the urinary bladder	Mannion and Susmano, 1971
52	F	Papillary carcinoma of the kidney	Bock and Hogrefe, 1972

Phenylbutazone

Phenylbutazone, a derivative of pyrazolone, was introduced in 1949 for the treatment of rheumatic diseases. It is also used as an analgesic, antipyretic, and anti-inflammatory agent.

In the liver it is almost completely metabolized to form oxyphenbutazone, a metabolite pharmacologically related to phenylbutazone, and hydroxyphenbutazone. Little is known about the final metabolic products. Oxyphenbutazone is metabolized slowly and tends to accumulate with repeated doses (Goodman and Gilman, 1970).

Among the most frequent side-effects of phenylbutazone are disorders of the gastrointestinal tract, reactivation of ulcers, and hematologic disturbances, including leukopenia, thrombocytopenia, agranulocytosis, and aplastic anemia (Chalmers and McCarthy, 1964; Goodman and Gilman, 1970).

Bean (1960) and Scheuer-Karpin (1965) drew attention to the possible leukemogenic effects of this drug.

We compiled 28 instances of a possible causal relation between phenylbutazone therapy and leukemia; these are shown in Table 11. Afflicted were 23 men and 25 women, aged 5–80 years at the time of diagnosis (average age of patients: 59 years). In most cases the drug had been prescribed for the treatment of inflammatory joint pains (Table 12); total doses ranged from 1–430 g (Table 13). Two patients received oxyphenbutazone. Phenylbutazone had been given for periods of 5 days to 12 years, interruptions not considered (treatment up to one year: 17 patients; 1–5 years: 15 patients; more than 5 years or for an unknown period: six patients). Latent periods ranged from a few months to 12 years. In Table 14 it can be seen that acute leukemias were observed most frequently (26 cases); in addition we found three subacute and six chronic leukemias. Three patients had other malignant diseases.

Two mechanisms are believed to be responsible for the action of phenylbutazone:

1. Phenylbutazone has direct leukemogenic effects, the mechanism of action of which is yet unknown. What is known is only the myelotoxicity of the drug. According to Dougan and Woodliff (1965) it is possible that the normal leukopoietic stimulus, acting on depressed bone marrow, may induce a leukemic stage. Thorpe (1964) considered leukemia to be the extreme form of phenylbutazone myelotoxicity. Hart (1964) also emphasized that an agent having such a broad spectrum of myelotoxicity might well be capable of inducing leukemia.
2. Phenylbutazone raises malignant hemoblastoses in rheumatics (Chatterjea, 1964; Lorenz and Gebert, 1968; Leavesley et al., 1969). Abatt and Lea (1958) speculated on a relationship between rheumatic disease and leukemia.

Dougan and Woodliff (1965) and Leavesley et al. (1969) found a history of phenylbutazone significantly more frequent in patients with acute leukemia than in those with chronic leukemia or other diseases. These authors, however, emphasized that among their patients there was no uniformity in age, sex, and basic disease.

According to Leavesley et al. (1969) only a few isolated cases of acute leukemia can be attributed to the use of phenylbutazone since among most of the cases other leukemogenic factors (i.e., irradiation) could not be excluded. It is also possible that in some cases the drug had been administered for the therapy of early symptoms of leukoses. Lorenz and Gebert emphasized that, especially in early stages, it might be difficult to distinguish between leukemic and rheumatic symptoms.

The average age of patients with acute leukemias (57 years) and the sex ratio (1.7 : 1 in favor of males) correspond to general observations (Begemann, 1970). The other forms of leukemia, because of the small number of cases, could not be evaluated.

The latent periods were strikingly short (< 2 years in 18 cases; 2–5 years in 16 cases). Those of other leukemias (i.e., induced by benzene or ionizing radiation) last at least 2 years and are even longer in most cases (Jensen and Roll, 1965).

The short latent periods raise doubts about a causal relationship between phenylbutazone therapy and leukemia (Fraumeni, 1967; Jensen and Roll, 1965).

References

Abatt, J. D., Lea, A. J.: Leukaemogens. Lancet 2, 880 (1958)

Bean, R. H. D.: Phenylbutazone and leukaemia: A possible association. Brit. Med. J. 2, 1552 (1960)

Begemann, H.: Klinische Hämatologie. Stuttgart: Georg Thieme, 1970

Cadman, E. F. B., Limont, W.: Phenylbutazone and leukaemia. Brit. Med. J. 1, 798 (1962)

Cast, I. P.: Phenylbutazone and leukaemia. Brit. Med. J. 2, 1569 (1961)

Chalmers, T. M., McCarthy, D. D.: Phenylbutazone therapy associated with leukaemia. Brit. Med. J. 1, 747 (1964)

Chatterjea, J. B.: Leukemia and Phenylbutazone. Brit. Med. J. 2, 875 (1964)

Dougan, L., Woodliff, H. J.: Acute leukaemia associated with phenylbutazone treatment: A review of the literature and report of a further case. Med. J. Aust. 1, 217 (1965)

Fraumeni, J. F.: Bone marrow depression induced by chloramphenicol or phenylbutazone. J. Amer. med. Ass. 201, 828 (1967)

Garrett, J. V.: Phenylbutazone and leukaemia. Brit. med. J. 1, 53 (1961)

Golding, J. R., Hamilton, M. G., Moody, H. E.: Monocytic leukemia and phenylbutazone. Brit. med. J. 1, 1673 (1965)

Goodman, L. S., Gilman, A.: The pharmacological basis of therapeutics. New York: The Macmillan Co. 1970

Hart, G. D.: Leukaemia and phenylbutazone. Brit. med. J. 2, 569 (1964)

Jensen, M. D., Roll, K.: Phenylbutazone and leukaemia. Acta med. scand. 176, 505 (1965)

Leavesley, G. M., Stenhouse, N. S., Dougan, L., Woodliff, H. J.: Phenylbutazone and leukaemia. Is there a relationship? Med. J. Aust. 56, 963 (1969)

Lorenz, K., Gebert, P.: Maligne Retikulose nach Phenylbutazon-Therapie. Kausalbeziehung oder zufälliges Zusammentreffen? Muench. med. Wschr. 110, 2283 (1968)

Peters, D., Sjöberg, S. G. (1965): cited by Fraumeni, J. F. (1967)

Scheuer-Karpin, R. (1965): cited by Lorenz, K., Gebert, P. (1968)

Sen, S., Siddique, K. K. H.: Phenylbutazone and leukaemia. Bull. Inst. postgrad. med. Educ. Res. 6, 23 (1964)

Stewart (1964): cited by Fraumeni, J. F. (1967)

Thorpe, G. J.: Leukaemia and phenylbutazone. Brit. med. J. 1, 1707 (1964)

Woodliff, H. J., Dougan, L.: Acute leukaemia associated with phenylbutazone treatment. Brit. med. J. 1, 744 (1964)

Table 11. Human leukemia associated with the use of phenylbutazone

Patient Age	Sex	Drug	Type of leukemia	Reference
69	M	Phenylbutazone	Chroic myeloid leukemia	Bean, 1960
67	M	Phenylbutazone	Acute lymphocytic leukemia	,,
70	M	Phenylbutazone	(Subacute?) lymphocytic leukemia	,,
80	M	Phenylbutazone	Chronic myeloid leukemia	,,
66	M	Phenylbutazone	Lymphocytic leukemia lymphosarcoma	,,

Table 11 (continued)

Patient Age Sex		Drug	Type of leukemia	Reference
63	M	Phenylbutazone	Chronic myeloid leukemia	„
59	F	Phenylbutazone	Chronic myeloid leukemia	Cast, 1961
64	F	Phenylbutazone	Acute lymphocytic leukemia	Garrett, 1961
71	M	Phenylbutazone	Acute stem cell leukemia	Cadman and Limont, 1962
52	F	Phenylbutazone	Monocytic myeloid leukemia	Chalmers and McCarthy, 1964
58	M	Phenylbutazone	Acute myeloid leukemia	Hart, 1964
29	M	Phenylbutazone	Acute myeloid leukemia	Sen and Siddique, 1964
56	F	Phenylbutazone	Acute stem cell leukemia	Thorpe, 1964
78	M	Phenylbutazone	Acute myeloid leukemia	Woodliff and Dougan, 1964
52	M	Phenylbutazone	Acute lymphocytic leukemia	„
80	M	Phenylbutazone	Acute myeloid leukemia	„
80	F	Phenylbutazone	Acute myeloid leukemia	„
58	F	Phenylbutazone	Acute myeloid leukemia	„
46	M	Phenylbutazone	Acute myeloid leukemia	Chatterjea, 1964
40	M	Phenylbutazone	Acute leukemia	Stewart, 1964
40	M	Oxyphenylbutazone	Acute myeloid leukemia	Peters and Sjöberg, 1965
47	M	Oxyphenylbutazone	Acute lymphocytic leukemia	„
57	F	Phenylbutazone	Acute monocytic leukemia	Golding et al., 1965
78	M	Phenylbutazone	Chronic myeloid leukemia	Jensen and Roll, 1965
67	M	Phenylbutazone	Acute myeloid leukemia	„
31	F	Phenylbutazone	Acute lymphocytic leukemia	„
58	F	Phenylbutazone	Acute lymphocytic leukemia	Dougan and Woodliff, 1965
60	M	Phenylbutazone	Chronic lymphatic leukemia	„
59	M	Phenylbutazone	Acute lymphocytic leukemia	„
61	F	Phenylbutazone	Subacute myeloid leukemia	Fraumeni, 1967
43	F	Phenylbutazone	Acute myeloid leukemia	„
77	F	Phenylbutazone	Acute myeloid leukemia	„
5	M	Phenylbutazone	Malignant reticulosis	Lorenz and Gebert, 1968
59	M	Phenylbutazone	Acute lymphocytic leukemia	Leavesley et al., 1969
51	F	Phenylbutazone	Acute myeloid leukemia	„
75	F	Phenylbutazone	Acute myeloid leukemia	„
64	M	Phenylbutazone	Erythroleukemia	„
57	M	Phenylbutazone	Acute myeloid leukemia	„

Table 12. Basic diseases of phenylbutazone-treated patients

Basic disease	Number of cases
Arthritis	14
Rheumatoid arthritis	10
Osteoarthritis	3
Osteoarthritis and spondylitis	1
Spondylitis	1
Rheumatism	1
Degenerative spondylitis	1
Cervical spondylitis	1
Sciatica	1
Lumbago	1
Backache	1
Bursitis	1
Phlebitis	1
No statement	1
Total number of cases	38

Table 13. Total doses of phenylbutazone applied to 38 patients

Total dose of phenylbutazone (g)	Number of cases
1– 10	13
11– 50	10
51–100	4
> 100	6
no statement	5
Total number of cases	38

Table 14. Patho-anatomy of phenylbutazone-related hemoblastoses

Type of leukosis	Number of cases
Acute myeloid leukemia	14
Acute lymphocytic leukemia	8
Acute stem cell leukemia	2
Acute monocytic leukemia	1
Acute leukemia	1

Table 14 (continued)

Type of leukosis	Number of cases
Subacute lymphocytic leukemia	1
Subacute monocytic myeloid leukemia	1
Subacute myeloid leukemia	1
Chronic myeloid leukemia	5
Chronic lymphocytic leukemia	1
Lymphosarcoma or lymphocytic leukemia	1
Malignant reticulosis	1
Erythroleukemia	1
Total number of cases	38

Diethylstilbestrol

Diethylstilbestrol is a synthetic estrogen that has the same pharmacologic effects as estradiol. It is also used for treating high risk pregnancy and preventing abortion.

Herbst et al. (1971a) and Greenwald et al. (1971) observed vaginal adenocarcinomas in 15 to 22-year-old women whose mothers had received diethylstilbestrol during pregnancy (survey and comment by Ulfelder, 1976). Up to 1972 the Registry of Clear-Cell Carcinoma of the United States had noted 64 cases of diethylstilbestrol-related carcinomas of the vagina and the cervix[5].

Table 15 contains a list of 23 cases. The age of the patients ranged from 7 to 22 years (average age: 17 years). In 20 cases the mothers had received diethylstilbestrol, and in two cases dienestrol was used. One mother had received both drugs. Treatment was started in the first trimester of gestation; 14 women were treated until or almost until the end of pregnancy; eight were treated for 3 months, and one was treated for 4 weeks. The doses varied considerably; a mean total dose therefore could not be determined. Some of the mothers had received graduated doses of 5–150 mg, others had received constant doses ranging from 25–100 mg/day. In 18 patients vaginal adenocarcinomas and in five patients cervical adenocarcinomas were formed. Histologic examination revealed 22 clear-cell adenocarcinomas and one endometrial carcinoma. Six patients died with tumors, 14 were without relapse 2 years after the operation.

---

[5] In the meantime this number has increased to more than 400 cases (Ivankovic, 1976, personal communication)

According to Herbst and Scully (1970) these tumors, which are characterized by glands and tubular forms consisting of cells rich in glycogen, originate from the Müllerian epithelium.

It was suggested that the carcinogen acted on the fetus during or after the development of the uterovaginal tract, which is formed by fusion of the Müllerian ducts. These are initially coated with a cylindric epithelium, which in the end of the first trimester of pregnancy is gradually replaced by squamous epithelium. It is possible that diethylstilbestrol disturbed this process since the gland tissue remained in the vagina. Increased endogenous estrogen stimulation during puberty might lead to malignant change of these structures by means of an adenosis.

This supposition was supported by Forsberg's finding (1972) that diethylstilbestrol and estradiol inhibited the transformation of the vaginal mucosa in experimental animals. In former studies (Forsberg, 1969), he had found that the changes leading to an adenosis occurred during the first and second month of life of the mice, i. e., the age of puberty. This development was prevented by castration of the animals before their second month of life. These findings demonstrate the role of the ovarian hormones in the formation of malignant vaginal changes.

In the present cases most of the stilbestrol-related tumors also occurred during puberty. Herbst et al. (1971 b, 1972) reported that in his patients benign adenomas were developed before the adenocarcinomas occurred. This indicates that the benign adenosis constituted a predisposing condition for the formation of an adenocarcinoma (Langmuir, 1971).

Carcinomas of the vagina are very rare and account for only 1–2% of all carcinomas of the female genital system. They occur normally after the 50th year of age (Herbst et al., 1971 b), predominantly in the form of squamous cell carcinomas; adenocarcinomas of the vagina are formed in only 5–6% of the cases (Herbst et al., 1971 b; Hill, 1973). Clear-cell carcinomas and adenocarcinomas of the cervix also occur predominantly in elderly women (Herbst et al., 1971 b).

A causal relation between intrauterine exposure to diethylstilbestrol and vaginal and cervical carcinomas is very likely, particularly since most patients were born at a time when stilbestrols came into widespread use (Folkmann, 1972). On the basis of these clinical observations further use of stilbestrols during pregnancy is rejected (Fetherston, 1972; Arzneimittelkommission der Deutschen Ärzteschaft, 1972). According to Uhlfelder et al. (1971), in Germany diethylstilbestrol was not used in pregnant women in such high doses as in the USA.

The present example demonstrates the significance of experimental findings indicating transplacental carcinogenic effects of chemical carcinogens (review by Ivankovic, 1975).

References

Arzneimittelkommission der Deutschen Ärzteschaft: Anwendung von Diäthylstilböstrol in der Schwangerschaft. Dtsch. Ärzteblatt *69*, 539 (1972)

Fetherston, W. C., Meyers, A., Speckhard, M. E.: Adenocarcinoma of the vagina in young women. The stilbestrol-adenosis-adenocarcinoma of the vagina syndrome. Wis. Med. J. *71*, 87 (1972)

Folkmann, J.: Transplacental carcinogenesis by stilbestrol. New Engl. J. Med. *285*, 404 (1971)

Forsberg, J. G. (1969): cited by Forsberg, J. G. (1972)

Forsberg, J. G.: Estrogen, vaginal cancer, and vaginal development. Amer. J. Obstet. Gynec. *113*, 83 (1972)

Greenwald, P., Barlow, J. J., Nasca, P. C., Burnett, W. S.: Vaginal cancer after maternal treatment with synthetic estrogens. New Engl. J. Med. *285*, 390 (1971)

Herbst, A. L., Kurman, R. J., Scully, R. E.: Vaginal and cervical abnormalities after exposure to stilbestrol in utero. Obstet. Gynec. *40*, 287 (1972)

Herbst, A. L., Scully, R. E.: Adenocarcinoma of the vagina in adolescence. A report of 7 cases including 6 clear-cell carcinomas (so-called mesonephromas). Cancer (Philad.) *25*, 745 (1970)

Herbst, A. L., Ulfelder, H., Poskanzer, D. C.: Adenocarcinoma of the vagina. Association of maternal stilbestrol therapy with tumor appearance in young women. New Engl. J. Med. *284*, 878 (1971 a)

Herbst, A. L., Ulfelder, H., Poskanzer, D. C.: Registry of clear-cell carcinoma of genital tract in young women. New Engl. J. Med. *285*, 407 (1971 b)

Hill, E. C.: Clear cell carcinoma of the cervix and vagina in young women. Amer. J. Obstet. Gynec. *116*, 470 (1973)

Ivankovic, S.: Praenatale Carcinogenese. In: Handbuch der allgemeinen Pathologie. Geschwülste, Tumors III, Altmann, H.-W., Büchner, F., Cottier, H., Grundmann, E., Holle, G., Letterer, E., Masshoff, W., Meesen, H., Roulet, F., Seifert, G., Siebert, G. (eds.) Berlin-Heidelberg-New York: Springer 1975 Vol. 6, part 7, pp. 941–1002

Kantor, H. I., Weinstein, S. A., Kaye, H. L.: Clear cell adenocarcinoma in young women. Obstet. Gynec. *41*, 443 (1973)

Kern, G.: Gynäkologie. Stuttgart: Georg Thieme 1973

Langmuir, A. D.: New environmental factor in congenital disease. New Engl. J. Med. *284*, 912 (1971)

Lewis, J. L., Nordquist, S. R. B., Richart, R. M.: Studies of nuclear DNA in vaginal adenosis and clear cell adenocarcinoma. Amer. J. Obstet. Gynec. *115*, 737 (1973)

Noller, K. L., Decker, D. G., Lanier, A. P., Kurland, L. T.: Clear-cell adenocarcinoma of the cervix after maternal treatment with synthetic estrogens. Mayo Clin. Proc. *47*, 629 (1972)

Ulfelder, H.: The stilbestrol-adenosis-carcinoma syndrome. Cancer (Philad.) *38*, 426 (1976)

Ulfelder, H., Poskanzer, D., Herbst, A. L.: Stilbestrol-adenosis-carcinoma syndrome geographic distribution. New Engl. J. Med. *285*, 691 (1971)

Vooijs, P. G., Ng, A. B. P., Wentz, W. B.: The detection of vaginal adenosis and clear cell carcinoma. Acta Cytol. (Baltimore) *17*, 59 (1973)

Table 15. Vaginal and cervical adenocarcinomas in young women prenatally exposed to diet ylstilbestrol and dienestrol

Age of Patient	Drug	Type of tumor	Reference
20	Diethylstilbestrol	Clear-cell adenocarcinoma of the vagina	Herbst et al., 1971 a
15	Diethylstilbestrol	Clear-cell adenocarcinoma of the vagina	,,
18	Diethylstilbestrol	Clear-cell adenocarcinoma of the vagina	,,
20	Diethylstilbestrol	Clear-cell adenocarcinoma of the vagina	,,
16	Diethylstilbestrol	Endometrioid carcinoma of the vagina	,,
19	Diethylstilbestrol	Clear-cell adenocarcinoma of the vagina	,,
22	Diethylstilbestrol	Clear-cell adenocarcinoma of the vagina	,,
15	Diethylstilbestrol	Clear-cell adenocarcinoma of the vagina	Greenwald et al., 1971
19	Diethylstilbestrol	Clear-cell adenocarcinoma of the vagina	,,
15	Diethylstilbestrol	Clear-cell adenocarcinoma of the vagina	,,
17	Dienestrol	Clear-cell adenocarcinoma of the vagina	,,
17	Diethylstilbestrol	Clear-cell adenocarcinoma	,,
19	Diethylstilbestrol	Clear-cell adenocarcinoma of the vagina	Fetherston, et al., 1972
7	Diethylstilbestrol	Clear-cell adenocarcinoma	Noller, et al., 1972
15	Diethylstilbestrol	Clear-cell adenocarcinoma of the cervix uteri	,,
19	Dienestrol	Clear-cell adenocarcinoma of the cervix uteri	,,
18	Diethylstilbestrol	Clear-cell adenocarcinoma of the vagina	Hill, 1973
16	Diethylstilbestrol	Clear-cell adenocarcinoma of the vagina	,,
17	Diethylstilbestrol	Clear-cell carcinoma of the cervix uteri	,,
14	Diethylstilbestrol	Clear-cell adenocarcinoma of the cervix uteri	Kantor et. al., 1973

Table 15 (continued)

Age of Patient	Drug	Type of tumor	Reference
16	Dienestrol and diethylstilbestrol	Clear-cell adenocarcinoma of the vagina	Kantor et al., 1973
11	Diethylstilbestrol	Clear-cell adenocarcinoma of the vagina	Vooijs et al., 1973
18	Diethylstilbestrol	Clear-cell adenocarcinoma	,,

Tar-Containing Ointments

Carcinogenic effects of tar for man have been known since Pott (1775) and Volkmann (1875) observed scrotal carcinomas following occupational exposure to tar. Because of their anti-inflammatory, antipruritic, and anti-infectious effects, both charcoal tar and pit-coal tar have been used in dermatology for a long time. Charcoal tar, which is a mixture of phenols, aromatic hydrocarbons, other aromatic compounds, and some low fatty acids (Moeller, 1961), is contained in ointments in a 2–10% concentration or is used in alcoholic solutions. Pit-coal tar (pix lithantracis, liquor carbonis detergens), which consists of phenol, naphthol, aromatic hydrocarbons, aniline, and pyridine and chinoline bases (Moeller, 1961), is used in ointments or pastes. It is applied predominantly in the treatment of psoriasis vulgaris. Possible side effects of epicutaneously absorbed tar preparations are renal lesions, sensibilization to light, and skin carcinomas, which are, however, extremely rare (Gottron, 1954, 1965).

Greither etal. (1967) compiled 13 cases of skin carcinoma associated with medical use of tar-containing drugs; these are shown in Table 16. The age of the patients was given in 9 cases and ranged from 27–85 years (average age: 54 years). Eleven patients were men, two were women. The tar preparations were used for the treatment of psoriasis vulgaris (four cases), eczema (three cases), pruritus (three cases), and cutaneous eruption (one case). In two cases the basic diseases were not described. One patient had used pine tar and three had used pit-coal tar. In the remaining cases no exact descriptions of the tar preparations used were given. Greither et al. (1967), however, assumed that pit-coal tar had been used in these cases also. Duration of treatment ranged from 3 months to 34 years, interruptions not considered. Tumor localizations were described as follows: scrotum (five cases), vulva, right groin, lower part of the abdomen, mammary gland, thigh, and skin of the head of the fibula (one case each). Latent periods of the tumors are not known. Seven tumors were histologically squamous cell carcinomas; in six cases the histologic diagnoses were not given. Formation of metastases was mentioned in only one case (squamous cell carcinoma of the vulva).

It is believed that the skin tumors were induced by polycyclic aromatic hydrocarbons contained in the tar (Kennaway, 1924; Hirohata et al., 1973). Pit-coal tar, some tar-containing ointments (carbon oil, carbopterine, lithantrol), and liquor picis carbonis were carcinogenic in experimental animals also (Yamagiwa and Ishikawa, 1918; Sternberg, 1923; Lipschütz, 1924; Berghoff, 1928; Berenblum, 1948).

Greither et al. (1967) attributed the formation of skin tumors to uncontrolled self-treatment for many years and emphasized that medically controlled, periodic use of small doses of tar preparations does not increase the risk of cancer.

The average age of the patients with tar-induced skin tumors (54 years), which corresponds to general observations, indicates long latent periods of the tumors. The increased incidence in men (11:2) has an external reason: pruritus genitalis was more often treated with tar preparations in the case of men than in women.

References

Alexander, I. O., Macrosson, K. I.: Squamous epithelioma probably due to tar ointment in a case with psoriasis. Brit. Med. J. 2, 1089 (1954)

Bauer, K. H.: Das Krebsproblem. Berlin-Heidelberg-Göttingen: Springer 1963

Berghoff, W.: Über Organveränderungen bei Mäusen nach Teerpinselung. Z. Krebsforsch. 26, 468 (1928)

Berenblum, I.: Liquor picis carbonis (B. P.). A carcinogenic agent. Brit. med. J. 2, 601 (1948)

Carli, G. (1958): cited by Greither, A., Gisbertz, C., Ippen, H. (1967)

De Jong, S. I., Meyer, J., Martineau, J. (1935): cited by Greither, A., Gisbertz, C., Ippen, H. (1967)

Gottron, H. A.: Gegenwartsfragen beim Hautkarzinom. Med. Klin. 49, 1553 (1954)

Gottron, H. A.: Karzinogene Wirkung teerhaltiger Salben? Med. Klin. 60, 1679 (1965)

Greither, A., Gisbertz, C., Ippen, H.: Teerbehandlung und Krebs. Z. Hautkr. 42, 631 (1967)

Henry, S. A., (1948): cited by Greither, A., Gisbertz, C., Ippen, H. (1967)

Hirohata, T., Masuda, Y., Horie, A., Kuratsune, M.: Carcinogenicity of tar-containing skin drugs: animal experiment and chemical analysis. Gann 64, 323 (1973)

Hodgson, G.: Epithelioma following the local treatment of pruritus ani with liquor picis carbonis. Brit. J. Derm. 60, 282 (1948)

Kennaway, E.: Brit. Med. J. 1, 564 (1924)

Kovtounovitch, G. P.: cited by Greither, A., Gisbertz, C., Ippen, H. (1967)

Lipschütz, B.: Z. Krebsforsch. *21*, 50 (1924)

Moeller, K. O.: Pharmakologie. Basel-Stuttgart: Schwabe & Co. 1961

Müller, S., Kierland, R.: Crude coal tar in dermatologic therapy. Proc. Mayo Clin. *39*, 275 (1964)

Opfer, K.: Karzinom auf Psoriasis durch Arsen- oder Teerbehandlung. Dermatologica *80*, 86 (1939)

Pott, P.: Chirurgical observations relative to the cataract, the polypos of the nose, the cancer of the scrotum, the different kinds of ruptures and the mortification of the toes and feet. London: 1775

Rook, A. J., Gresham, G. A., Davis, R. A.: Squamous epithelioma possibly induced by the therapeutic application of tar. Brit. J. Cancer *10*, 17 (1956)

Sonneck, H. J., Koch, H.: Skrotalkarzinom nach langjähriger unkontrollierter Teerbehandlung eines Ekzems. Derm. Wschr. *148*, 40 (1963)

Sternberg, A.: Z. Kerbsforsch. *20*, 420 (1923)

Veiel, F.: Teerkrebs beim Menschen. Arch. Derm. Syph. (Berl.) *148*, 142 (1924)

Volkmann, S. (1875): cited by Bauer, K. H. (1963)

Yamagiwa, K., Ishikawa, K. (1918): cited by Bauer, K. H. (1963)

Table 16. Skin carcinomas following therapeutic application of tar

Patient Age	Sex	Type of tumor	Reference
68	M	Carcinoma of the scrotum	Veiel, 1924
40	F	Malignant abdominal ulceration, tumorous changes of the breast and in the abdomen	Kovtounovitch, 1927
85	M	Spinocellular carcinoma of the thigh	De Jong et al., 1935
57	M	Spinocellular carcinoma of the scrotum	Opfer, 1939
−	M	Skin carcinoma	Henry, 1948
−	M	Skin carcinoma	„
−	M	Spinocellular carcinoma of the scrotum	Hodgson, 1948
32	M	Spinocellular carcinoma of a psoriatic efflorescence in the area of the head of the fibula	Alexander and Macrosson, 1954
60	M	Squamous cell carcinoma of the inguinal region	Rook et al., 1956
27	M	Carcinoma of the scrotum	Carli, 1958
63	F	Spinocellular carcinoma of the vulva	„
68	M	Squamous cell carcinoma of the scrotum	Sonneck and Koch, 1963
−	M	Skin carcinoma	Müller and Kierland, 1964

Examples of Iatrogenic Carcinogenesis in the Field of Surgery and its Marginal Areas

Foreign-Body and Scar Sarcomas

Several plastic implants, including polyethylene, polyvinylchloride, bakelite, polyamide, polystyrol, and polyurethane, led to the formation of sarcomas in rats, mice, hamsters, and dogs. The same applies to metals (gold, silver, and platinum) and other materials, such as ivory, window glass, gum, quartz, silk, and parchment paper. The low tumor yield and the particularly long latent periods of the sarcomas, however, indicate only weak carcinogenic effects of these materials (for references see Ott et al., 1963).

From human pathology, there are only a few known cases of sarcoma associated with foreign-body implants. These are shown in Table 17. Table 18 contains a compilation of 20 cases of sarcoma in operation scars. Both the foreign-body and the scar sarcomas were formed on the ground of a scarred, avascular fibroplasia of the connective tissue. Histologic examination of the foreign-body sarcomas showed one Ewing's sarcoma, one alveolar sarcoma, one chondrosarcoma, one rhabdomyosarcoma, and two fibrosarcomas. Histologically, most of the scar sarcomas were fibrosarcomas.

According to Ott (1970), both foreign-body and scar sarcomas develop on the basis of avascular scar tissue, which, its pathogenesis notwithstanding, is considered a presarcomatosis. In foreign-body sarcomas the quantity of scar tissue formed depends on size, porosity, and surface of the implant.

A survey of the experimental findings shows the following:

1. The larger the surface of an implant, the larger the scar capsule and accordingly the greater the risk of developing a foreign-body sarcoma (Nothdurft, 1955; Stout and Eirich, 1955; Hueper, 1961; Ott et al., 1963; Ott, 1970).

2. The more porous the foreign body, the smaller the fibrous capsule and accordingly the less the risk of cancer (Contzen, 1963; Ott et al., 1963), except when the implant has fibrosing effects itself (i. e. porous cancellous tissue; Ott and Jansen, 1966).

3. Concave foreign-body surfaces increase the incidence of sarcomas (Nothdurft, 1961), an effect that was attributed to an increased scarring (Contzen, 1963; Contzen et al., 1967).

4. Hardness and elasticity of the implant do not influence the quantity of scar tissue formed (Nothdurft, 1961; Contzen et al., 1967).

5. In contrast to the earlier hypothesis of Druckrey and Schmähl (1952) and Fitzhugh (1953), chemical carcinogens contained in the implants, except for some known carcinogens (i.e., zinc in metal alloys), have no direct effects on the formation of sarcomas (Mohr and Nothdurft, 1958; Ott, 1970).

6. Latent periods of the foreign-body sarcomas range from one third to one half of the mean life expectancy of the animals (Ott et al., 1963).

7. The risk of cancer was hardly reduced by removal of the foreign body when it was implanted for at least one third of the expected lifetime and the surrounding connective tissue capsule was not removed together with the implant.

8. At the onset of the healing process, granulation tissue, unspecific and rich in cells, is formed, which later becomes cell deficient and rich in collagen, forming a fibrous capsule around the implant. After 6–8 months, proliferating cells are observed that are characterized by polymorphic cell nuclei and numerous mitoses. The rapidly growing sarcomas develop from these proliferating cells (Mohr and Nothdurft, 1958; Ott et al., 1963).

Because of the small number of clinical observations a causal relation between the insertion of foreign bodies and the formation of sarcomas can only be suspected. It is difficult to assess whether the implant itself, traumatic impacts, preneoplastic changes, or irradiation account for the development of the sarcomas or whether syncarcinogenic effects of all these factors have to be considered. These questions will probably be answered within the next few decades, because in recent years larger foreign bodies have been used also (Ott, 1970). The latent periods of foreign-body sarcomas are rated at 30 to 60 years (Hueper, 1964; Druckrey, 1960).

Considering the large number of surgical interventions during the past years, the risk of developing a scar sarcoma is extremely small, although we assume that a number of cases have not been registered.

References

Abruzzini, P., Vecchione, F.: Fibrosarcoma gigante della parete poracica successivo ad intervento di resezione polmonare. Chir. Thorac. *14*, 31 (1961)

Betzler, H. J., Leonhardt, J.: Sarkom in einer Laparotomienarbe. Z. Krebsforsch. *63*, 118 (1959)

Bürkle de la Camp, H. (1958): cited by Struppler, V. (1959)

Burns, W. A., Kanhouwa, S., Tillman, L., Saini, N., Herrmann, B.: Fibrosarcoma occurring at the site of a plastic vascular graft. Cancer (Philad.) *29*, 66 (1972)

Capaldi, A.: Ein Sarkom am Amputationsstumpf nach 20 Jahren. Muench. med. Wschr. 55, 512 (1930)

Contzen, H.: Die lokale Gewebereaktion auf implantierte Kunststoffe in Abhängigkeit von deren Form. Langenbecks Arch. klin. Chir. 304, 922 (1963)

Contzen, H., Straumann, F., Paschke, E.: Grundlagen der Alloplastik mit Metallen und Kunststoffen. Stuttgart: Georg Thieme 1967

Druckrey, H.: Die Entstehung von Krebs. Monatliche Kurse ärztl. Fortbildung 10, 199 (1960)

Druckrey, H., Schmähl, D.: Cancerogene Wirkung von Kunststoff-Folien. Z. Naturforsch. Teil B 7, 353 (1952)

Fitzhugh, A. F.: Malignant tumors and high polymers. Science 118, 783 (1953)

Fuchs, G., Stegemann, H., Eger, W.: Der transplantierte Knochenspan und seine Qualität nach partieller und vollständiger Enteiweißung bei erhaltener anorganischer Substanz. Langenbecks Arch. klin. Chir. 303, 240 (1963)

Grueter and Höbler (1959): cited by Ott, G. (1970)

Hamant, A., Cornil, L., Mosinger, A.: Sarcome fibroblastique de la cuisse developpe sur une cicatrice operatoire. Ann. anat. path. 7, 373 (1930)

Heller, E. L., Sieber, W. K.: Fibrosarcoma. Surgery 27, 539 (1950)

Hueper, W. C.: Surgical use of polyurethane plastics questioned on basis of animal experiments. Amer. J. clin. Path. 34, 328 (1961)

Hueper, W. C. (1964): cited by Ott, G. (1970)

Ju, D. M. C.: Fibrosarcoma arising in surgical scars. Plast. reconstr. Surg. 38, 429 (1966)

McDougall, A.: Malignant tumours at site of bone plating. J. Bone Jt. Surg. 38, 709 (1956)

Mohr, H. J., Nothdurft, H.: Bindegewebskapseln um subcutan eingeheilte Fremdkörper und ihre Entartung zu Sarkomen. Klin. Wschr. 36, 493 (1958)

Nolte, D.: Über Sarkomentstehung nach Kriegsverletzung. Mschr. Unfallheilk. 69, 124 (1966)

Nothdurft, H.: Über die Sarkomauslösung durch Fremdkörperimplantationen bei Ratten in Abhängigkeit von der Form der Implantate. Naturwissenschaften 42, 106 (1955)

Nothdurft, H.: Sarkomerzeugung bei Ratten durch implantierte Fremdkörper. Ther. Mh. Mannheim: Boehringer 8, 262 (1961)

Ott, G.: Fremdkörpersarkome. In: Experimentelle Medizin, Pathologie und Klinik. Leuthardt, F., Schoen, R., Schwiegk, H., Studer, A., Zollinger, H. U. (eds.) Berlin – Heidelberg – New York: Siringer, 1970

Ott, G., Jansen, H. H.: Sarkomentstehung nach Homio- und Heterotransplantation von Knochen. Langenbecks Arch. klin. Chir. 314, 1 (1966)

Ott, G., Vollmar, J., Hieronymi, G.: Krebsgefährdung nach Implantation von Kunststoffen. Langenbecks Arch. klin. Chir. 302, 608 (1963)

Pack, G. T., Ariel, J. M.: Fibrosarcoma of the soft somatic tissue. Surg. Gynec. Obstet. 98, 675 (1954)

Stout (1957): cited by Ott, G. (1970)

Stout and Eirich (1955): cited by Ott, G. (1970)

Struppler, V.: Sarkome nach Knochennagelung. Mschr. Unfallheilk. 62, 121 (1959)

Warren and Sommer (1936): cited by Ott, G.(1970)

Zülch (1953): cited by Ott, G. (1970)

Table 17. Foreign-body sarcomas

Localization Trauma Foreign-body	Latency period (years)	Type of tumor	Reference
Humerus, fracture, bone plating	30	Ewing's sarcoma	McDaugal, 1956
Thigh, fracture, marrow nailing	4	Alveolar sarcoma	Bürkle de la Camp, 1958
Neck of the femur, fracture, nail excision after 2 months	6	Chondrosarcoma	Struppler, 1959
Tibia, fracture, transplantation of the chip of the tibia	21	Rhabdomyosarcoma	Nolte, 1966
Head of the tibia, operative implantation of spongiosa from the os ileum of the father because of lacking erythropoiesis	10	Fibrosarcoma	Ott, 1970
Toe, rupture of a femoral artery, teflon-dracon prosthesis	10	Fibrosarcoma	Burns et al., 1972

Table 18. Sarcomas at the site of operation scars

Site of tumor	Latency period (years)	Type of tumor	Reference
Lower arm, amputation scar	20	Sarcoma	Capaldi, 1930
Thigh, fibromectomy scar	1	Fibrosarcoma (relapse of a primary sarcoma?)	Hamant et al., 1930
Scar of the neck after incision because of lymphadenitis	50	Fibrosarcoma	Warren and Sommer, 1936
Mastectomy scar		Fibrosarcoma	Heller and Sieber, 1950
Laparotomy scar after excision of a transverse colon carcinoma	2	Fibrosarcoma	Pack and Ariel, 1954
Cranial dura after trepanation	9	Fibrosarcoma	Zülch, 1953
1. Laparotomy scar,	2	Fibrosarcoma	Stout, 1957
2. Axillary scar after incision,	20	Fibrosarcoma	„
3. Scar of the finger after incision of an abscess,	15	Fibrosarcoma	„
4. Appendectomy scar	18	Fibrosarcoma	„
Laparotomy scar after oversewing of an ulcus perforans, cholecystectomy	19[a] 4[b]	– Fibrosarcoma	„ Betzler and Leonhardt, 1959
Tendon of the triceps surae after achillotenotomy	4	Proliferating fibrosarcoma	Grueter and Höbler, 1959
Thoracotomy scar (periosteum of the 5th rib) after lobectomy for tuberculosis	2	Fibrosarcoma	Abruzzini and Vecchione, 1961
Periosteum of the 4th rib after pneumolysis	2	Fibrosarcoma	Fuchs et al., 1963
1. Appendectomy scar,	26	Fibrosarcoma	Ju, 1966
2. scar of the sternum after removal of a fibrosarcoma and a neurofibrosarcoma,	11	Fibrosarcoma	„
3. Back (excision of a leiomyoma	10	Fibrosarcoma	„

Table 18 (continued)

Site of tumor	Latency period (years)	Type of tumor	Reference
4. Inguinal region (herniotomy)	15	Fibrosarcoma	Ju, 1966
1. Scar of the breast after mastitis incision	15	Fibrosarcoma	Ott, 1970
2. Laparotomy scar after bridenileus	5	Fibrosarcoma	„

[a] After oversewing of an ulcer.
[b] After cholecystectomy.

Skin Tumors at the Site of Smallpox Vaccination Scars

Complications following smallpox vaccination include abnormal courses of the local reaction, such as postvaccinal ulcers, vaccinia generalisata, eczema vaccinatum, and skin tumors at the site of vaccination, which are, however, rare.

In a survey of the literature we found 54 cases of postvaccinal tumors (Table 19). Those afflicted included 42 men and 12 women, aged 15–81 years at the time of diagnosis (average age: 52 years). In one case the sex was not given. Latent periods of the tumors ranged from a few weeks to 80 years. In 51 patients the tumors occurred at the site of vaccination, in one patient tumors developed on both arms, one patient had tumors in 3 different vaccination scars on both arms, and in two patients the tumors occurred on the shoulder and knuckle. Histologic examination of the tumors, which were 2–5 cm in size, dark brown to black and occasionally melanotic, revealed 34 basal cell carcinomas, 13 malignant melanomas, eight squamous cell carcinomas, one fibrosarcoma, one dermatofibrosarcoma protuberans, and one intra-epidermic basal cell epithelioma. Metastases were formed in only three patients.

The following mechanisms of skin cancer induction by smallpox vaccination have been considered:

1. The vaccinial virus might be the direct cause of the skin tumors (Auger, 1943; Marmelzat, 1968; Geiser, 1970): In rabbits the application of smallpox viruses led to the formation of fibrosarcomas and myxomas (Marmelzat, 1968); co-carcinogenic factors were probably not involved. Duran-Reynals (1957), Duran-Reynals and

Stanley (1961), and Siegler and Duran-Reynals (1962) showed experimentally that the vaccinial virus intensified the action of chemical carcinogens. Ilie (1963) increased the rate of appearance of reticulosarcomas and other tumors from 7%, with single application of Ehrlich ascites cells, to 53% with simultaneous application of vaccinial viruses.

2. Another possibility is malignant degeneration of the vaccination scar, which occasionally occurs in scars due to burning or irradiation, lupus vulgaris, and syphilitic gums (Marmelzat, 1968; Archampong and Clark, 1970; Geiser, 1970).

3. In some cases exposure of the vaccination scar to sunlight (UV light) was thought to be the cause of the malignant change (Dorsey et al., 1960; Reed and Wilson-Jones, 1968; Geiser, 1970).

4. In a few patients probably predisposing factors were involved. Geiser (1970) observed a total of 16 patients who had additional basal cell carcinomas, squamous and basal cell carcinomas, squamous cell carcinomas, keratoacanthomas, and another malignant skin neoplasm that was not classified further.

In the present cases the ages of patients were below those observed in other skin tumors. (Basal cell and squamous cell carcinomas commonly occur between the 60th and 80th year of age; malignant melanomas are most frequently observed between the 40th and 70th year of age.) The sex ratio (3.5:1) showed a distinct preference for men. The three most common types of malignant skin tumors, squamous cell carcinomas, basal cell carcinomas, and melanomas, were represented. The tumors were formed in the deltoid region, which in most cases was the site of vaccination. It is therefore unlikely that the present observations were coincidental.

References

Archampong, E. Q., Clark, C. G.: Fibrosarcoma at the site and immediately following small-pox vaccination. Brit. J. Surg. 57, 937 (1970)

Auger, C.: Cancer sur tatouge et cancer sur cicatrice de vaccination antivariolique. Laval méd. 8, 300 (1943)

Bazex, A., Dupre, A., Christol, B. (1968): cited by Geiser, J.-D. (1970)

Coetzee, T. (1964): cited by Geiser, J.-D. (1970)

Dorsey, C. S., Marmelzat, W. L., Levan, N.: Skin cancer in smallpox vaccination scars: A report of five cases. Calif. Med. 92, 353 (1960)

Duran-Reynals, F.: Preliminary studies on the development of neoplasia in the skin of mice painted with methylcholanthrene and injected with cortisone and vaccina virus. Ann. N. Y. Acad. Sci. 68, 430 (1957)

Duran-Reynals, M. G., Stanley, B.: Vaccina dermal infection and methylcholanthrene in cortisone-treated mice. Science 134, 1984 (1961)

Geiser, J.-D.: Tumeurs malignes, complications de cicatrices de vaccination antivariolique. Praxis 59, 1158 (1970)

Goncalves, J. C.: Malignant change in smallpox vaccination scars. Arch. Derm. 93, 229 (1966)

Ilie, B. (1963): cited by Marmelzat, W. L. (1968)

Many, P., Lapeyre, J., Boutet, B., Taillard, P. (1969): cited by Geiser, J.-D. (1970)

Marmelzat, W. L.: Malignant tumours in smallpox vaccination scars. A report of 24 cases. Arch. Derm. *97*, 400 (1968)

Marmelzat, W. L., Hirsch, P., Martel, S.: Malignant melanomas in smallpox vaccination scars. Arch. Derm. *89*, 823 (1964)

Porter, D., Earle, J.: Jadassohn tumour arising at smallpox vaccination site. Brit. J. Derm. *86*, 177 (1972)

Rea, E.: Squamous carcinoma on vaccination scar. S. Afr. med. J. *30*, 499 (1956)

Reed, W. B., Wilson-Jones, W.: Malignant tumours as a late complication of vaccination. Arch. Derm. *98*, 132 (1968)

Riley, K. A.: Basal cell epithelioma in smallpox vaccination scar. Report of a case. Arch. Derm. *101*, 416 (1970)

Siegler, R., Duran-Reynals, M. L.: Observations on the pathogenesis of experimental skin tumors: A study on the mechanisms by which papillomas develop. J. nat. Cancer Inst. *29*, 653 (1962)

Washington, L. P., Jacobs, P. H.: Basal cell cancer in a vaccination scar. Cutis *8*, 557 (1971)

Weary, P. E. (1967): cited by Geiser, J.-D. (1970)

Zelickson, A. S.: Basal cell epithelioma at site and following smallpox vaccination. Arch. Derm. *98*, 35 (1968)

Table 19. Skin tumors at the site of smallpox vaccination scars

Patient Age	Sex	Type of tumor	Reference
64	M	Squamous cell carcinoma	Auger, 1943
40	F	Squamous cell carcinoma	Rea, 1956
54	M	Basal cell carcinoma	Dorsey et al., 1960
48	M	Basal cell carcinoma	,,
48	–	Basal cell carcinoma	,,
44	M	Basal cell carcinoma	,,
53	F	Basal cell carcinoma	,,
47	M	Dermatofibrosarcoma protuberans	Coetzee, 1964
31	F	Malignant melanoma	Marmelzat et al., 1964
42	M	Malignant melanoma	,,
34	F	Malignant melanoma	,,
68	M	Malignant melanoma	,,
76	F	Malignant melanoma	,,
66	M	Malignant melanoma	,,
60	M	Basal cell carcinoma	Goncalves, 1966
33	M	Basal cell carcinoma	Weary, 1967
81	M	Basal cell carcinoma	Bazex et al., 1968
63	M	Basal cell carcinoma	Marmelzat, 1968
50	M	Basal cell carcinoma	,,

Table 19 (continued)

Patient Age	Sex	Type of tumor	Reference
46	M	Basal-cell carcinoma	Marmelzat, 1968
60	M	Basal cell carcinoma	,,
56	M	Basal cell carcinoma	,,
45	M	Basal cell carcinoma	,,
30	M	Basal cell carcinoma	,,
(38)	–	(Basal cell carcinoma)	,,
37	M	Basal cell carcinoma	,,
35	F	Basal cell carcinoma	,,
15	F	Basal cell carcinoma	,,
58	M	Basal cell carcinoma	,,
53	M	Basal cell carcinoma	,,
61	M	Basal cell carcinoma	,,
49	F	Squamous cell carcinoma	,,
60	M	Squamous cell carcinoma	,,
34	M	Squamous cell carcinoma	,,
69	M	Squamous cell carcinoma in situ	,,
66	M	Squamous cell carcinoma in situ	,,
38	M	Malignant melanoma	,,
58	M	Malignant melanoma	,,
49	F	Malignant melanoma	,,
32	M	Malignant melanoma	,,
43	M	Malignant melanoma	,,
53	M	Malignant melanoma	,,
41	M	Basal cell epithelioma	Reed and Wilson-Jones, 1968
58	M	Basal cell carcinoma	,,
70	M	Basal cell carcinoma	,,
80	F	Basal cell carcinoma	,,
29	M	Squamous cell carcinoma	,,
58	F	Malignant melanoma	,,
51	M	Basal cell carcinoma	Zelickson, 1968
72	M	Basal cell carcinoma	Many et al., 1969
61	M	Basal cell carcinoma	Geiser, 1970
74	M	Basal cell carcinoma	,,
44	F	Fibrosarcoma	Archampong and Clark, 1970
64	M	Basal cell carcinoma	Riley, 1970
53	M	Basal cell carcinoma	Washington and Jacobs, 1971
74	M	Intraepidermic basal cell epithelioma	Porter and Earle, 1972

Primary Carcinoma of the Operated Stomach

The term refers to those carcinomas that were diagnosed after an operation of a benign gastric ulcer. By definition, an interval of 5 years should elapse between the operation and the manifestation of the tumor.

Since K. H. Bauer's first survey (1951), in which he compiled 26 cases of carcinoma of the operated stomach, the number of observations has considerably increased[6]. We found 327 cases in the world literature; these are shown in Table 20. The age of the patients (271 men; 54 women) ranged from 34 to 83 years (average age: 59 years). In two cases the sex was not given.

Indications for the operation were ulcus ventriculi (160 cases), ulcus duodeni (127 cases), ulcus pylori (eight cases), ulcus ventriculi et duodeni (six cases), and ulcus duodeni et jejuni (two cases). In the remaining 24 cases the site of ulcer was not given. In most cases gastric resection and gastroenterostomy had been used. Free intervals between the operation and the occurrence of tumor ranged from 5 to 47 years (average free interval: 19 years). The site of tumor was described in 206 cases; tumors of the anastomotic area were developed in only 55 cases (Table 21). (Boeckl and Lill (1963), in a study of 299 patients with primary carcinoma of the operated stomach, found 54 carcinomas in the anastomotic area, 32 in the cardia and fundus of the stomach, and 75 in the remaining parts of the stomach; in the remaining 138 cases the site of tumor was not described). Histologically, 92 tumors were adenocarcinomas, 14 were solid carcinomas, ten were scirrhous carcinomas; 20 tumors were diagnosed as "carcinoma."

In 153 cases, the course of the disease could be traced: 22 patients survived more than 2 years, 87 died within the first 6 months after the time of diagnosis, 30 died between the 6th and 12th month, and 14 died between the 12th and 24th months after the diagnosis. Metastases were formed most frequently in the regional lymph nodes and in the liver.

According to Saegesser and Jämes (1972) the size of the primary

---

[6] Reviews by Prinz (1938), Freedman and Berne (1953), Denk and Salzer (1957), Berkowitz et al. (1959), Liavaag (1962), Eggeling (1963), Prévot (1963), Capper and Johnson (1964), Grahame (1965), Vitek et al. (1966), Albrecht et al. (1966), Pygott and Shah (1968), Morgenstern (1968), Becker (1969), Kobayashi et al. (1970), Howanietz and Wense (1971), Stalsberg and Taksdal (1971), Ottenjann (1973), Morgenstern et al. (1973), Kowalewski (1973), Taksdal and Stalsberg (1973), Moulinier et al. (1973), Griffel et al. (1974), Nicholls (1974), Rehner et al. (1974), Mitschke and Dahm (1975)

carcinoma does not influence the incidence of metastases. Both the macroscopic and microscopic appearance of the tumors did not differ from that of other primary epithelial neoplasms (Debray et al., 1958; Daroczi and Metzl, 1959; Wanke, 1971). Most investigators observed polypous forms, which tended to form hemorrhages, hard pruniform tumors, and ulcerous forms in the size of a palm (Kyrle and Wild, 1952; Pack and Banner, 1958; Daroczi and Metzl, 1959). Penetration into the pancreas, transverse mesocolon, retroperitoneum, and liver was observed more frequently (Kyrle and Wild, 1952; Daroczi and Metzl, 1959; Kühlmayer, 1964) than passing to the small intestine (Kyrle and Wild, 1952; Debray et al., 1958). The anastomotic carcinomas tended to spread circularly; stenoses were formed early (Debray et al., 1958; Wanke, 1971).

The site of ulcer obviously plays an important role in the formation of these tumors. Helsingen and Hillestad (1956) found a threefold increased incidence of gastric carcinomas in patients with ulcus ventriculi compared to those with ulcus duodeni. The latter group of patients showed the same incidence of gastric carcinomas as non-operated patients. Griesser and Schmidt (1964) observed 13–16% gastric tumors in patients with ulcus ventriculi compared to 6.25% in those with ulcus duodeni.

Huber (1953) considered the following explanations of the formation of tumors in the operated stomach:

1. A malignant process was present that was not recognized at the time of the operation; the tumor was therefore a relapse.
2. An ulcer, which was not resected by the operation or which developed later, degenerated malignantly.
3. The operation favored the formation of tumor.

In reference to 1: Primary wrong diagnoses can be excluded with respect to the definition (5 years' free interval).

In reference to 2: According to Finsterer (1923) and Anschütz and Wanke (1931), malignant degeneration of benign ulcers may occur after insufficient gastric resection or gastroenterostomy.

In reference to 3: This theory was investigated by several authors. Postoperative chronic gastritis might be a pathogenic factor (Konjetzny, 1938; Bandmann, 1953; Heinzel and Laqua, 1954; Neumann, 1957; Gremmel et al., 1959; Heinzel et al., 1960; Huber and Deutschmann, 1960; Kootz, 1967). Anschütz and Wanke (1931), Usland (1935), and von Oppolzer (1938) even supposed that each chronic gastritis (in the resected or non-operated stomach) may degenerate malignantly. In the present cases gastritis accompanying chronic ulcers, anastomotic and stump gastritis would have to be considered (Neumann, 1957; Huber and Deutschmann, 1960). Konjetzny (1938) and Boller (1952) found chronic alterations of the anastomotic mucosa in histologic and gastroscopic examinations. According to Evers (1962), Koelsch (1962), Heinkel et al. (1964), and Seifert et al. (1966), the mucosa of the resected stomach was altered pathologically only a few years after the operation. Seifert et al. (1966) found predominantly superficial gastritis during the first 5 years after the operation; after an additional 5 years atrophic gastritis predominated, and after 20 years the glands of the gastric fundus had disappeared.

Anacidity in the resected stomach was also supposed to be a tumor-favoring condition (Boller, 1952; Katsch and Pickert, 1953; Debray et al., 1958). In patients with pernicious agastric anemia the incidence of gastric cancer was found to be increased (Heinzel and Laqua, 1954). Further factors are unphysiologic junction of two different epithelial tissues at the anastomotic site (Krause, 1958; Vargha and Repasy, 1963), mechanic overexertion of the anastomotic region (Bandmann, 1953), and unphysiologic chemical irritation by the reflux of alkaline juices (Bandmann, 1953; Pack and Banner, 1958; Lenzenweger, 1959). Griesser and Schmidt (1964) emphasized the significance of the operative method. In a study of patients with operated ulcus ventriculi they found 15–16% gastric tumors after gastrojejunostomy (with or without gastric resection) compared to 7% after gastroduodenostomy.

Dahm and Werner (1973) showed experimentally that N-methyl-nitroso-nitro-guanidine-induced tumors developed earlier in the area of gastroenteral anastomoses and concluded that the anastomoses provided dispositional changes that favored an increased and accelerated tumor growth in this part of the stomach.

The incidence of gastric cancer of the operated stomach was studied by several authors who reported contradictory results. These are shown in Table 22. Hilbe et al. (1968) attributed these differences to the different methods of investigation used. They stated the following:

1. Some authors (Becker and Freund, 1964; Gregl and Wiedenmann, 1966; Kootz, 1967) compared the number of gastric carcinomas with that of stump carcinomas observed during a certain period of time. These findings show only the relative frequency of stump carcinomas among te gastric carcinomas altogether; however, they do not permit a conclusion about the general frequency of stump carcinomas.

2. Comparison of the number of stump carcinomas diagnosed during a certain period of time with the number of ulcer resections carried out during the same period (Heinzel et al., 1960; Becker and Freund, 1964; Schreiber et al., 1964; Kootz, 1967; Kronberger and Hafner, 1968) showed too low values since in the choice of the case material investigated, the 20 years' free interval was not considered, and the number of gastric resections has increased in recent years.

3. In the postoperative examination of patients with an operated ulcer, wrong diagnoses or disregard of the free interval is possible. Helsingen and Hillestad (1956) and

Griesser and Schmidt (1964) found that the rate of gastric carcinomas in patients operated for an ulcus ventriculi was increased compared to that of nonoperated individuals.

4. In the evaluation of post-mortem findings, which indicated a predominance of primary carcinomas of the operated stomach (Kühlmayer and Rokitansky, 1954; Hebold, 1958), a positive selection has to be considered. Hilbe et al. (1968), in post-mortem examinations carried out from 1945 to 1964, found 8.2% carcinomas in patients with gastric operations compared to 5.4% in nonoperated individuals.

Final clarification of these questions might be obtained only by a comparison of the incidence of gastric cancer in patients with gastric resection with that of nonoperated individuals, and autopsies in order to exclude wrong diagnoses.

The average age of the patients described here corresponded to general observations in gastric tumors (Kühlmeyer and Rokitansky, 1954; Strauss, 1957; Griesser and Schmidt, 1964). The sex ratio (5:1 in favor of males) may be due to the increased rate of ulcers and accordingly of gastric resections in men. The average free interval between the operation and the manifestation of a tumor (19 years) is in accordance with the findings of most investigators (Boeckl and Lill, 1963; Gerstenberg et al., 1965; Hilbe et al., 1968).

References

Albrecht, A., Gerstenberg, E., Krentz, K., Voth, H.: Das Magenstumpfkarzinom: Diagnose und Differentialdiagnose. Radiologie *6*, 353 (1966)

Anschütz, W., Wanke, R. (1931): cited by Pirner, F. (1953)

Asseburg, U., Wiegandt, F., Manitz, G.: Zur Problematik primärer Karzinome im operierten Magen. Med. Welt *24*, 1508 (1973)

Bandmann, F.: Über die Carcinombildung am Gastro-Entero-Anastomosenring. Bruns Beitr. klin. Chir. *186*, 210 (1953)

Bauer, K. H. (1951): cited by Heinzel, J., Hess, H., Laqua, H. (1960)

Beatson, G. T.: Carcinoma of the stomach after gastrojejunostomy. Brit. Med. J. *1*, 15 (1926)

Becker, Th., Freund, E.: Magencarcinom und Ulcuschirurgie. Zbl. Chir. *89*, 455 (1964)

Becker, V.: Pathologische Anatomie des resezierten Magens. In: Magenoperation und Magenoperierter. Bartelheimer, H., Maurer, J. H., Schreiber, H. W. (eds.) Berlin: de Gruyter 1969

Berkowitz, D., Cooney, P., Bralow, S. P.: Carcinoma of the stomach appearing after previous gastric surgery for benign ulcer disease. Gastroenterology *36*, 691 (1959)

Berry, T. J., Lee, T. C., Coffey, R. J.: Carcinoma arising in the gastric stump following gastric resection for benign ulceration. Amer. J. Surg. *25*, 353 (1959)

Bessot, M., Gillot, C., Laurent, J., Vicari, F., Arbogast, J., Heuly, F.: Trois cas de cancer apparu sur éstomacs opérés. Sem. Hôp. Paris *46*, 1217 (1970)

Boeckl, O., Lill, H.: Über das Magenstumpfkarzinom. Muench. med. Wschr. *105*, 615 (1963)

Boller, R.: Der operierte Magen. Wien. klin. Wschr. *203* (1952)

Burian, J.: Der primäre Krebs in dem wegen Gastroduodenalgeschwür resezierten Magen. Zbl. Chir. *85*, 2223 (1960)

Capper, W., Johnson, H. D.: Vagotomy and carcinoma of the stomach. Lancet *2*, 1063 (1964)

Cote, R., Dockerty, M. B., Cain, J. C.: Cancer of the stomach after gastric resection for peptic ulcer. Surg. Gynec. Obstet. *107*, 200 (1958)

Dahm, K., Werner, B.: Anastomosenkarzinom im resezierten Magen der Ratte nach Gabe von N-Methyl-N'nitrosoguanidin. Dtsch. med. Wschr. *52*, 2486 (1973)

Daroczi, G., Metzl, J.: Über das sogenannte "Magenstumpfkarzinom." Bruns Beitr. klin. Chir. *198*, 401 (1959)

Debray, Ch., Bouvry, M., Roches, Ph.: Über das Stumpfkarzinom nach Magenresektion wegen Ulcus. Schweiz. med. Wschr. *88*, 631 (1958)

Denk, H., Salzer, G.: 21 Jahre Ulcuschirurgie an der Klinik Denk in Wien 1933–1954. II. Teil: Die Frage der Karzinomgefährdung des Ulcuskranken und Magenresezierten. Gastroenterologia (Basel) *88*, 94 (1957)

Eggeling, W.: Das Karzinom im Ulcus-Resektionsmagen. Fortschr. Med. *81*, 167 (1963)

Evers, Ch.: Saugbiopsie am operierten Magen. Med. Klin. *57*, 1080 (1962)

Feldman, F., Seaman, W. B.: Primary gastric stump cancer. Amer. J. Roentgenol. *115*, 257 (1972)

Finsterer, H. (1923): cited by Neumann, W. D. (1957)

Fontaine, R., Warter, P., Weill, F.: Etude radiologique de sept cancers développés sur des moignons de gastrectomie pour ulcère. J. Radiol. Electrol. *43*, 465 (1962)

Freedman, M. A., Berne, C. J.: Gastric carcinoma of gastrojejunal stomach. Gastroenterology *40*, 210 (1954)

Gerstenberg, E., Albrecht, A., Krentz, K., Voth, H.: Das Magenstumpfkarzinom: eine Spätkomplikation des operierten Magens? Dtsch. med. Wschr. *90*, 2185 (1965)

Grahame, E. W.: Vagotomy and carcinoma of the stomach. Lancet *1*, 109 (1965)

Gregl, A., Wiedenmann, R.-W.: Karzinome im Restmagen. Bruns Beitr. klin. Chir. *213*, 177 (1966)

Gremmel, H., Schulte-Brinkmann, W., Vieten, H.: Karzinome nach Magenoperationen wegen nicht blastomatöser Erkrankungen. Fortschr. Roentgenstr. *90*, 342 (1959)

Griffel, B., Engleberg, M., Reiss, R.: Multiple polypoid cystic gastritis in old gastroenteric stomach. Arch. Path. (Chicago) *97*, 316 (1974)

Griesser, G., Schmidt, H.: Statistische Erhebungen über die Häufigkeit des Karzinoms nach Magenoperationen wegen eines Geschwürleidens. Med. Welt *15*, 1836 (1964)

Guivarc'h, M., Marquand, J., Mouchet, A.: Cancer et moignon gastrique. A propos de 30 observations. Ann. Chir. *22*, 337 (1968)

Hebold, G.: Das Carcinom im Restmagen. Med. Klin. *53*, 1813 (1958)

Heinkel, K., Henning, N., Parpoulas, S., Landgraf, J., Elster, K. (1964): cited by Hilbe, G., Salzer, G. M., Hussl, H., Kutschera, H. (1968)

Heinzel, J., Hess, H., Laqua, H.: Karzinombildung in Mägen, die wegen Ulcus ventriculi bzw. duodeni reseziert wurden. Bruns Beitr. klin. Chir. *201*, 156 (1960)

Heinzel, J., Laqua, H.: Magencarcinoma nach früherer Resektion wegen Ulcus ventriculi bzw. duodeni. Langenbecks Arch. klin. Chir. *278*, 87 (1954)

Helsingen, N., Hillestad, L.: Cancer development in the gastric stump after partial gastrectomy for ulcer. Ann. Surg. *143*, 173 (1956)

Hilbe, G., Salzer, G. M., Hussl, H., Kutschera, H.: Die Carcinomgefährdung des Resektionsmagens. Langenbecks Arch. klin. Chir. *323*, 142 (1968)

Hollender, L.-F., Muller, P., Sini, F., Santizo, G.: La cancérisation secondaire du moignon restant après résection gastrique subtotale pour ulcère. Arch. Mal. Appar. dig. *55*, 625 (1966)

Howanietz, L., Wense, G.: Das Magenstumpfkarzinom, eine Spätkomplikation nach Ulkusresektion. Zbl. Chir. *96*, 437 (1971)

Huber, P.: Über Karzinomentwicklung im operierten Magen. Bruns Beitr. klin. Chir. *186*, 317 (1953)

Huber, P., Deutschmann, W.: Über Krebsentwicklung im operierten Magen. Landarzt *36*, 1287 (1960)

Justin-Besancon, L., Deuil, R., Grivaux, M., Etienne, J. P., Guerre, J., Mundler, B.: Les cancers du moignon gastrique. Sem. Hôp. Paris *41*, 3079 (1965)

Katsch, G., Pickert, H.: Handbuch der inneren Medizin. Berlin-Göttingen-Heidelberg: Springer 1953, Vol. III

Kny, W.: Beitrag zum Magenstumpfkarzinom. Bruns Beitr. klin. Chir. *204*, 106 (1962)

Kobayashi, S., Prolla, J. C., Kirsner, J. B.: Late gastric carcinoma developing after surgery for benign conditions. Amer. J. Dig. Dis. *15*, 905 (1970)

Koelsch, K. A.: Saugbioptische Untersuchungen bei Magenoperierten. Muench. med. Wschr. *104*, 2384 (1962)

Konjetzny, G. E.: Der Magenkrebs. Stuttgart: Ehmke 1938

Kootz, F.: Das Stumpfkarzinom nach Operation eines benignen Magenleidens. Bruns Beitr. klin. Chir. *215*, 275 (1967)

Kowalewski, K.: Relationship between vagotomy, peptic ulcer, and gastric adenocarcinoma in rats fed 2,7-diacetylaminofluorene. Canad. J. Surg. *16*, 210 (1973)

Krause, U.: Late prognosis after partial gastrectomy for ulcer. Acta Chir. Scand. *114*, 341 (1958)

Kronberger, L., Hafner, H.: Über das "primäre Stumpfkarzinom" nach Ulcusresection. Chirurg. *39*, 118 (1968)

Kühlmayer, R.: Zum Problem des Magenstumpfkarzinoms. Muench. med. Wschr. *76*, 293 (1964)

Kühlmayer, R., Rokitansky, O.: Das Magenstumpfkarzinom als Spätproblem der Ulcuschirurgie. Langenbecks Arch. klin. Chir. *278*, 361 (1954)

Kyrle, P., Wild, H.: Über Magenstumpfkarzinome. Zbl. Chir. *77*, 1481 (1952)

Lagache, G., Vankemmel, M.: Les cancers du moignon gastriques après gastrectomie pour ulcère. Ann. Chir. *20*, 618 (1966)

Lenzenberger, F.: Über Karzinomentwicklung am Magen nach Gastroenterostomie und Resektion wegen benigner Erkrankungen. Krebsarzt *14*, 99 (1959)

Liavaag, K.: Cancer development in gastric stump after partial gastrectomy for peptic ulcer. Ann. Surg. *155*, 103 (1962)

Mitschke, H., Dahm, K.: Definition, Häufigkeit, Pathogenese, Pathologische Anatomie. In: Das Karzinom im operierten Magen. Dahm, K., Rehner, M. (eds.) Stuttgart: Georg Thieme 1975

Morgenstern, L.: Vagotomy, gastroenterostomy and experimental gastric cancer. Arch. Surg. (Chicago) *96*, 920 (1968)

Morgenstern, L., Yamakawa, T., Seltzer, D.: Carcinoma of the gastric stump. Amer. J. Surg. *125*, 29 (1973)

Mouchet, A., Marquand, J., Garcin, J. P., Boury, G.: Les cancers du moignon gastrique après gastrectomie pour ulcère. (Etude critique à propos 5 cas). Ann. Chir. *17*, 137 (1963)

Moulinier, B., Lamgert, R., Grenier-Boley, P., Ruet, D., Truchot, R.: Etude éndoscopique du cancer gastrique après gastrectomie pour ulcère. Ann. Gastroent. Hepat. *9*, 359 (1973)

Neumann, W.-D.: Zur Carcinomentwicklung im Restmagen nach Resektion wegen Ulcus duodeni oder ventriculi. Chirurg *28*, 15 (1957)

77

Nicholls, J. C.: Carcinoma of the stomach following partial gastrectomy for benign gastro-duodenal lesions. Brit. J. Surg. *61*, 244 (1974)

v. Oppolzer, R. (1938): cited by Pirner, F. (1953)

Ottenjann, R.: Der operierte Magen und seine Folgezustände. In: Klinische Gastro-enterologie. Demling, L. (ed.) Stuttgart: Thieme 1973, Vol. I, pp. 263–272

Pack, G. T., Banner, R.: The late development of gastric cancer after gastroenterostomy and gastrectomy for peptic ulcer and benign pyloric stenosis. Surgery *44*, 1024 (1958)

Pirner, F.: Zum Thema Magenstumpfkarzinom. Zbl. Chir. *78*, 646 (1953)

Prévot, R.: Die Röntgendiagnostik des operierten Magens. Dtsch. med. Wschr. *88*, 942 (1963)

Prinz, H.: Über Krebsbildungen im Gastroenterostomosenring und deren Bedeutung für die Lehre von der Krebsentstehung im Magen. Langenbecks Arch. klin. Chir. *191*, 140 (1938)

Prots, R. A., Dragstedt, L. R.: Carcinoma of stomach 40 years after gastroenterostomy. Ohio St. Med. J. *67*, 521 (1971)

Pygott, F., Shah, V. L.: Gastric cancer associated with gastroenterostomy and partial gastrectomy. Gut *9*, 117 (1968)

Rehner, M., Soehendra, N., Eichfuß, H. P., Dahm, K., Eckert, P., Mitschke, H.: Frühkarzinome im (Billroth II-) Resektionsmagen. Dtsch. med. Wschr. *99*, 533 (1974)

Saegesser, F., Jämes, D.: Cancer of the gastric stump after gastrectomy (Billroth II principle) for ulcer. Cancer (Philad.) *29*, 1150 (1972)

Schäfer, R.: Zum Problem des Magenstumpfkarzinoms nach Geschwürresektion. Zbl. Chir. *87*, 1584 (1962)

Schreiber, W. H., Bernhard, A., Kuss, B.: Über das Karzinom im Magenstumpf. Zbl. Chir. *89*, 577 (1964)

Seifert, E., Dittrich, H., Erd, W.: Gastrobioptische Untersuchungen am Resektions-magen. Med. Welt *17*, 38 (1966)

Stalsberg, H., Taksdal, S.: Stomach cancer following gastric surgery for benign con-ditions. Lancet *2*, 1175 (1971)

Strauss, K.-J. (1957): cited by Gerstenberg, E., Albrecht, A., Krentz, K., Voth, H. (1965)

Taksdal, S., Stalsberg, H.: Histology of gastric carcinoma occurring after gastric surgery for benign conditions. Cancer (Philad.) *32*, 162 (1973)

Usland, O. (1935): cited by Pirner, F. (1953)

Vargha, J., Repasy, S.: Über primäre Anastomosenkarzinome nach Resektion des Magens nach Billroth I. Bruns Beitr. klin. Chir. *206*, 373 (1963)

Vitek, J., Vrubel, F., Zejda, V.: Primary carcinoma of the gastric stump following resection for ulcer. Radiologe *6*, 359 (1966)

Wanke, M.: Das Primärkarzinom im operierten Magen. In: Spezielle pathologische Anatomie. Doerr, W., Seifert, G., Uehlinger, E. (eds.). Berlin-Heidelberg-New York: Springer 1971, Vol. II, pp. 117–1044

Table 20. Primary carcinomas of the operated stomach

Patient Age	Sex	Operation for	Operative method/diagnosis	Reference
42	F	U. duodeni	Gastroenterostomy	Kyrle and Wild, 1952
51	M	U. duodeni	Billroth's second method	,,
42	F	U. ventriculi	Billroth's second method	,,
44	F	U. ventriculi	Billroth's second method	,,
62	F	—	Billroth's second method	,,
62	M	U. ventriculi	Billroth's second method	,,
45	M	U. ventriculi	Billroth's second method	,,
50	M	U. ventriculi	Billroth's second method	,,
56	M	U. ventriculi	Billroth's second method	,,
61	M	Ulcer scar of the pylorus	Gastroenterostomy	Huber, 1953
53	M	—	Transverse resection	,,
70	F	U. ventriculi	Billroth's second method	,,
54	M	U. ventricul	Billroth's second method	,,
75	F	U. duodeni	Billroth's second method	,,
63	F	U. duodeni	Billroth's second method	,,
64	M	—	Billroth's second method	,,
50	M	U. duodeni	Billroth's second method	,,
56	F	U. ventriculi et duodeni	Billroth's second method	,,
66	F	U. duodeni	Gastroenterostomy	Bandmann, 1953
60	—	U. ventriculi	Billroth's second method	,,
61	F	U. ventriculi	Billroth's second method	,,
68	—	U. ventriculi	Billroth's second method	,,
52	M	U. duodeni	Billroth's second method	Pirner, 1953
45	M	U. duodeni	Billroth's second method	Heinzel and Laqua, 1954
34	F	U. ventriculi	Billroth's second method	,,
50	M	U. duodeni	Gastric resection	,,
52	M	U. duodeni	Billroth's first method	,,
66	M	U. ventriculi et duodeni	Gastric resection	Helsingen and Hillestad, 1956
66	M	U. duodeni	Gastric resection	,,
60	M	U. ventriculi	Gastric resection	,,
69	M	U. ventriculi	Gastric resection	,,
55	M	U. ventriculi	Gastric resection	,,
60	F	U. ventriculi	Gastric resection	,,
46	M	U. ventriculi	Gastric resection	,,

Table 20 (continued)

Patient Age Sex		Operation for	Operative method/diagnosis	Reference
44	F	U. ventriculi	Gastric resection	Helsingen and Hillestad, 1956
59	F	U. ventriculi	Gastric resection	,,
73	M	U. ventriculi	Gastric resection	,,
71	M	U. ventriculi	Gastric resection	,,
49	M	U. duodeni	Gastric resection	Neumann, 1957
51	F	U. ventriculi	Gastroenterostomy	Pack and Banner, 1958
44	M	U. ventriculi	Gastroenterostomy	,,
61	M	U. duodeni	Gastroenterostomy	,,
72	F	U. duodeni	Gastroenterostomy	,,
52	F	U. duodeni	Gastroenterostomy	,,
67	M	U. duodeni	Gastroenterostomy	,,
43	F	U. duodeni	Gastroenterostomy	,,
35	M	U. duodeni	Gastroenterostomy	,,
61	M	U. duodeni	Gastroenterostomy	,,
59	F	U. ventriculi	Gastroenterostomy	,,
71	M	U. duodeni	Gastroenterostomy	,,
56	F	U. duodeni	Subtotal gastrectomy	,,
57	M	U. duodeni	Gastroenterostomy	,,
69	M	U. ventriculi	Atresia of the pylorus	Cote et al., 1958
41	M	U. duodeni	Subtotal gastrectomy	,,
53	M	U. ventriculi	Gastric resection	,,
56	M	U. ventriculi	Gastric resection	,,
62	F	U. ventriculi et duodeni	Gastric resection	,,
52	M	U. ventriculi	Gastroenterostomy	,,
67	M	U. duodeni	Gastroenterostomy	,,
65	M	U. ventriculi	Gastroenterostomy	,,
69	F	U. ventriculi	Gastroenterostomy.	,,
74	M	U. ventriculi	Gastroenterostomy	,,
50	M	U. ventriculi	Gastroenterostomy	,,
51	M	U. ventriculi	Gastroenterostomy	,,
52	M	—	Billroth's second method	Hebold, 1958
69	M	—	Billroth's second method	,,
66	M	—	Billroth's second method	,,
69	F	—	Billroth's first method	,,
70	M	—	Billroth's second method	,,
72	M	—	Billroth's second method	,,

Table 20 (continued)

Patient Age	Sex	Operation for	Operative method/diagnosis	Reference
76	M	–	Billroth's second method	Hebold, 1958
78	F	–	Billroth's second method	,,
57	M	U. ventriculi	Billroth's second method	Berry et al., 1959
56	M	U. ventriculi	Gastroenterostomy	,,
59	M	U. duodeni	Billroth's second method	Lenzenweger, 1959
51	M	U. ventriculi	Billroth's second method	,,
67	M	U. duodeni	Billroth's second method	,,
58	F	U. duodeni	Gastroenterostomy	,,
62	M	U. ventriculi	Billroth's second method	,,
45	M	U. duodeni	Gastroenterostomy	,,
58	M	U. ventriculi	Billroth's first method	,,
62	M	U. duodeni	Gastroenterostomy	,,
71	F	U. duodeni	Gastroenterostomy	,,
52	M	U. ventriculi	Billroth's second method	,,
55	M	U. ventriculi	Billroth's second method	,,
51	M	U. duodeni	Gastroenterostomy	,,
59	M	U. duodeni	Billroth's second method	,,
64	M	U. ventriculi	Billroth's second method	,,
47	M	–	Billroth's second method	,,
41	M	U. duodeni	Billroth's first method	,,
57	M	U. ventriculi	Billroth's second method	,,
58	M	U. ventriculi	Billroth's second method	,,
58	F	U. ventriculi	Billroth's first method	,,
60	M	U. duodeni	Gastroenterostomy	,,
74	M	U. duodeni	Billroth's second method	,,
57	M	U. duodeni	Billroth's second method	,,
53	M	U. ventriculi	Billroth's second method	,,
58	M	–	Billroth's second method	,,
56	F	U. duodeni	Billroth's second method	,,
72	F	U. ventriculi	Gastroenterostomy	,,
75	M	–	Billroth's first method	,,
69	M	U. duodeni	Billroth's second method	,,
62	F	U. duodeni	Billroth's second method	,,
54	M	U. ventriculi et duodeni	Billroth's second method	,,
67	M	U. duodeni	Billroth's second method	,,
70	F	U. duodeni	Billroth's second method	,,

Table 20 (continued)

Patient Age	Sex	Operation for	Operative method/diagnosis	Reference
66	M	U. ventriculi	Billroth's second method	Lenzenweger, 1959
43	F	U. duodeni	Gastroenterostomy	"
60	M	U. ventriculi	Gastroenterostomy	Gremmel et al., 1959
49	M	U. ventriculi	Gastroenterostomy	"
39	M	U. ventriculi	Gastroenterostomy	"
54	M	U. ventriculi	Billroth's first method	"
48	M	U. ventriculi	Billroth's first method	"
61	F	U. ventriculi	Billroth's second method	"
54	M	U. ventriculi	Billroth's second method	"
51	M	U. ventriculi	Billroth's second method	"
50	M	U. ventriculi	Billroth's second method	"
50	M	U. duodeni	Billroth's second method	"
62	M	U. duodeni	Billroth's second method	"
56	M	U. duodeni	Billroth's second method	"
46	M	U. duodeni	Billroth's first method	Daroczi and Metzl, 1959
59	M	U. duodeni	Billroth's second method	"
59	M	U. duodeni	Billroth's second method	"
56	M	U. duodeni	Billroth's second method	"
66	M	U. duodeni	Billroth's second method	"
52	M	−	Billroth's second method	Burian, 1960
48	M	U. ventriculi	Billroth's second method	"
64	M	U. ventriculi	Billroth's second method	"
60	M	U. duodeni	Billroth's second method	"
55	M	U. duodeni	Billroth's second method	"
56	M	U. ventriculi	Billroth's second method	"
64	M	−	Billroth's second method	"
58	M	U. ventriculi	Billroth's second method	"
49	M	U. duodeni	Billroth's second method	"
62	M	U. ventriculi	Billroth's second method	"
63	F	U. ventriculi	Billroth's second method	"
61	M	U. ventriculi	Billroth's second method	"
55	M	U. duodeni	Billroth's second method	"
50	M	U. duodeni	Billroth's second method	"
55	M	U. duodeni	Billroth's second method	"
67	M	U. ventriculi	Billroth's second method	"
66	F	U. ventriculi	Billroth's second method	"
55	M	U. ventriculi	Billroth's second method	"

Table 20 (continued)

Patient Age	Sex	Operation for	Operative method/diagnosis	Reference
57	M	U. duodeni	Billroth's second method	Burian, 1960
80	M	–	Billroth's second method	,,
52	M	U. ventriculi	Billroth's second method	,,
60	M	U. ventriculi	Billroth's first method	,,
66	M	U. duodeni	Billroth's second method	Heinzel et al., 1960
66	M	U. duodeni	Billroth's first method	,,
48	M	U. ventriculi	Billroth's second method	,,
55	M	U. duodeni	Billroth's second method	,,
54	M	U. ventriculi	Billroth's second method	,,
62	M	U. duodeni	Billroth's second method	,,
69	M	U. ventriculi	Billroth's second method	,,
56	F	U. ventriculi	Billroth's second method	,,
67	M	U. duodeni	Billroth's second method	,,
49	M	U. ventriculi	Billroth's second method	,,
77	M	U. ventriculi	Billroth's second method	,,
46	M	U. ventriculi	Billroth's second method	,,
55	M	U. ventriculi	Billroth's second method	,,
67	M	U. ventriculi	Billroth's second method	,,
54	M	U. ventriculi	Billroth's second method	,,
75	M	U. ventriculi	Billroth's second method	,,
73	M	U. ventriculi	Billroth's second method	Schäfer, 1962
70	F	U. ventriculi	Billroth's first method	,,
50	M	U. duodeni	Billroth's second method	,,
71	M	U. ventriculi	Atresia of the pylorus	Fontaine et al., 1962
68	M	U. ventriculi	Billroth's second method	,,
45	M	U. ventriculi	Billroth's second method	,,
73	M	U. ventriculi	Atresia of the pylorus	,,
69	M	U. duodeni	Atresia of the pylorus	,,
60	M	U. duodeni	Atresia of the pylorus	,,
64	M	U. duodeni et ventriculi	Atresia of the pylorus	Mouchet et al., 1963
61	M	U. pylori	Gastric resection	,,
44	M	u. duodeni	Atresia of the pylorus	,,
54	M	U. ventriculi	Subtotal gastrectomy	,,
59	F	U. ventriculi	Subtotal gastrectomy	,,
67	F	U. duodeni	Gastroenterostomy	Kny, 1962
75	M	U. duodeni	Gastric resection	,,

Table 20 (continued)

Patient Age	Sex	Operation for	Operative method/diagnosis	Reference
61	M	U. duodeni	Billroth's first method	Vargha and Repasy, 1963
51	M	U. ventriculi	Gastroenterostomy	Becker and Freund, 1964
70	M	U. ventriculi	Gastroenterostomy	,,
34	M	U. ventriculi	Gastroenterostomy	,,
71	F	U. ventriculi	Gastroenterostomy	,,
65	M	U. ventriculi	Gastroenterostomy	,,
55	M	U. ventriculi	Gastroenterostomy	,,
57	F	U. ventriculi	Gastroenterostomy	,,
66	F	U. ventriculi	Gastroenterostomy	,,
56	M	U. ventriculi	Billroth's first method	,,
61	M	U. ventriculi	Billroth's first method	,,
52	F	U. ventriculi	Billroth's first method	,,
63	M	U. ventriculi	Billroth's first method	,,
75	F	U. ventriculi	Billroth's first method	,,
56	M	U. ventriculi	Billroth's second method	,,
38	M	U. ventriculi	Billroth's second method	,,
54	F	U. ventriculi	Billroth's second method	,,
50	M	U. ventriculi	Billroth's second method	,,
54	F	U. ventriculi	Billroth's second method	,,
58	M	U. ventriculi	Billroth's second method	,,
67	M	U. ventriculi	Billroth's second method	,,
46	M	U. ventriculi	Billroth's second method	,,
58	M	U. ventriculi	Billroth's second method	,,
50	M	U. ventriculi	Billroth's second method	,,
77	M	U. ventriculi	Billroth's second method	,,
71	F	U. ventriculi	Billroth's second method	,,
61	M	U. ventriculi	Billroth's second method	,,
65	M	U. ventriculi	Billroth's second method	,,
72	M	U. ventriculi	Billroth's second method	,,
57	M	U. duodeni	Billroth's second method	,,
63	M	U. duodeni	$^2/_3$ resection	Justin-Besancon et al., 1965
75	F	–	$^2/_3$ resection	,,
53	M	U. ventriculi	Gastrectomy	,,
58	M	U. ventriculi	Gastrectomy	,,
56	M	U. duodeni	$^2/_3$ resection	,,
59	M	–	Gastrectomy	,,
70	M	U. pylori	$^2/_3$ resection	,,

Table 20 (continued)

Patient Age	Sex	Operation for	Operative method/diagnosis	Reference
55	M	U. prepyl.	$^2/_3$ resection	Justin-Besancon et al., 1965
54	M	U. ventriculi	Billroth's second method	Gerstenberg et al., 1965
53	M	U. ventriculi	Billroth's second method	,,
64	M	U. ventricui	Billroth's second method	,,
70	M	U. ventriculi	Billroth's second method	,,
68	M	U. ventriculi	Billroth's second method	,,
58	M	U. ventriculi	Billroth's second method	,,
64	F	U. ventricli	Billroth's second method	,,
38	M	U. ventriculi	Billroth's second method	,,
53	M	U. duodeni	Billroth's second method	,,
60	M	U. duodeni	Billroth's second method	,,
68	M	U. duodeni	Billroth's second method	,,
52	M	U. duodeni	Billroth's second method	,,
54	M	U. duodeni	Billroth's second method	,,
59	M	U. duodeni	Billroth's second method	,,
52	M	U. duodeni	Billroth's second method	,,
60	M	U. duodeni	Billroth's second method	,,
66	M	. duodeni	Gastroenterostomy	,,
60	M	U. duodeni	Gasroenterostomy	,,
64	F	U. ventriculi	Gastroenterostomy	,,
58	M	U. ventriculi	Gastroenterostomy	,,
42	M	U. ventriculi	Atresia of the pylorus	Lagache and Vankemmel, 1966
63	M	U. ventriculi	$^2/_3$ resection	,,
55	F	U. duodeni	Gastrectomy	,,
57	M	U. duodeni	Gastrectomy	,,
44	M	U. duodeni	$^2/_3$ resection	,,
53	M	—	$^2/_3$ resection	,,
49	M	U. duodeni	$^2/_3$ resection	,,
65	M	U. duodeni	Gastroenterostomy	,,
64	M	U. ventriculi	$^2/_3$ resection	,,
56	M	U. pylori	Atresia of the pylorus	Hollender et al., 1966
61	M	U. ventriculi	Atresia of the pylorus	,,
71	M	U. pylori	Péan's method	,,
53	M	U. duodeni	Atresia of the pylorus	,,
60	M	U. duodeni	Atresia of the pylorus	,,

Table 20 (continued)

Patient Age	Sex	Operation for	Operative method/diagnosis	Reference
43	M	U. pylori	Atresia of the pylorus	Hollender et al., 1966
67	M	U. ventriculi	Atresia of the pylorus	„
57	M	U. duodeni	Atresia of the pylorus	„
69	M	U. ventriculi	Atresia of the pylorus	„
58	M	U. ventriculi	Atresia of the pylorus	„
63	M	U. duodeni	Finsterer's method	„
62	F	U. ventriculi	Atresia of the pylorus	„
59	M	U. ventriculi	Atresia of the pylorus	„
65	M	U. ventriculi	Finsterer's method	„
50	M	U. duodeni	Billroth's second method	Kootz, 1967
54	M	U. ventriculi	Billroth's second method	„
67	M	Ulcus	Billroth's second method	„
66	F	U. ventriculi	Gastroenterostomy	„
50	M	U. duodeni	Billroth's second method	„
68	M	U. duodeni	Billroth's second method	„
52	M	U. duodeni	Billroth's second method	„
60	M	U. ventriculi	Billroth's second method	„
50	M	U. ventriculi	Billroth's second method	„
60	M	U. duodeni	Gastroenterostomy	„
52	M	U. ventriculi	Billroth's second method	„
65	M	U. ventriculi	Gastroenterostomy	„
69	M	U. ventriculi	Gastroenterostomy	„
69	M	U. duodeni	Gastroenterostomy	„
60	M	Ulcus	Billroth's second method	„
63	M	U. ventriculi	Gastroenterostomy	„
49	M	U. ventriculi	Billroth's second method	„
67	M	U. duodeni	Billroth's second method	„
47	M	U. duodeni	Billroth's second method	„
60	M	U. ventriculi	Atresia of the pylorus	Guivarc'h et al., 1968
56	M	U. ventriculi	Finsterer's method	„
68	M	U. ventriculi	Gastroenterostomy and atresia of the pylorus	„
58	M	U. duodeni	Finsterer's method	„
73	M	U. ventriculi	Atresia of the pylorus	„
64	M	U. duodeni et ventriculi	Atresia of the pylorus	„
61	M	U. pylori	Atresia of the pylorus	„

86

Table 20 (continued)

Patient Age	Sex	Operation for	Operative method/diagnosis	Reference
44	M	U. duodeni	Atresia of the pylorus	Guivarc'h et al., 1968
65	M	Ulcus	Gastrectomy	Bessot et al., 1970
73	M	U. ventriculi	Gastrectomy	Prots and Dragstedt, 1971
54	M	U. duodeni	Billroth's second method	Feldman and Seamen, 1972
79	M	U. ventriculi	Billroth's second method	,,
64	F	U. ventriculi	Billroth's second method	,,
79	M	U. ventriculi	Billroth's second method	,,
54	M	U. duodeni	Billroth's second method	,,
57	M	U. duodeni	Billroth's second method	,,
66	M	U. duodeni	Billroth's second method	,,
56	M	U. duodeni	Billroth's second method	,,
72	M	U. ventriculi	Billroth's second method	,,
70	M	U. duodeni	Billroth's second method	,,
62	F	U. ventriculi	Billroth's second method	,,
71	M	Ulcus	Billroth's second method	,,
66	M	U. duodeni	Billroth's second method	Saegesser and Jämes, 1972
64	F	U. duodeni	Billroth's second method	,,
51	M	U. duodeni	Billroth's second method	,,
65	M	U. duodeni	Billroth's second method	,,
58	M	U. ventriculi	Billroth's second method	,,
54	M	U. duodeni	Billroth's wecond method	,,
57	M	U. ventriculi	Billroth's second method	,,
58	M	U. duodeni	Billroth's second method	,,
73	M	U. duodeni	Billroth's second method	,,
57	M	U. duodeni et jejuni	Gastroenterostomy and Billroth's second method	,,
83	M	U. duodeni	Billroth's second method	,,
58	M	U. duodeni	Billroth's second method	,,
60	M	U. duodeni et jejuni	Gastroenterostomy and Billroth's second method	,,
74	M	U. duodeni	Billroth's second method	,,
73	M	U. duodeni	Billroth's second method	,,
59	M	U. duodeni	Billroth's second method	,,
74	M	U. ventriculi	Billroth's second method	,,

Table 20 (continued)

Patient Age	Sex	Operation for	Operative method/diagnosis	Reference
64	M	U. duodeni	Billroth's second method	Saegesser and Jämes, 1972
60	M	U. ventriculi	Billroth's second method	Asseburg et al., 1973
53	M	U. ventriculi	Gastroenterostomy and Billroth's second method	,,
61	M	U. duodeni	Billroth's second method	,,
58	M	U. duodeni	Billroth's second method	,,
61	F	U. duodeni	Billroth's second method	,,
48	M	U. duodeni	Billroth's second method	,,
59	M	U. ventriculi	Billroth's second method	,,
61	M	U. duodeni	Billroth's second method	,,

Table 21. Localizations of primary carcinomas of the operated stomach (327 cases)

Site of tumor	Number of cases
Anastomosis	55
Anastomosis and stump	6
Cardia	28
Fundus	7
Corpus	20
Lesser curvature	14
Greater curvature	5
Posterior gastric wall	6
Anterior gastric wall	2
Antrum	14
Diffuse carcinomas	36
Multiple carcinomas	13
No statement	121
Total number of cases	327

Table 22. Percentage of primary carcinomas of the operated stomach. Results of several authors

Investigator(s)	Percentage of primary carcinomas of the operated stomach
Griesser and Schmidt	13.3–14.10[a]
	6.25[b]
Kühlmayer and Rokitansky	10.60
Hilbe et al.	8.20
Krause	8.00
Helsingen and Hillestad	5.00
Heinzel et al.	1.35
Becker and Freund	1.17
Kronberger and Hafner	1.10

[a] after operation of an ulcus ventriculi
[b] after operation of an ulcus duodeni

The Stewart-Treves Syndrome

Lymphangiosarcomas occurring with lymphedema following radical mastectomy were first described by Stewart and Treves (1948). Numerous case reports have since confirmed the etiology, clinical symptomatology, microanatomy, and prognosis of the Stewart-Treves syndrome. The term has been extended to include angiosarcomas in congenital, posttraumatic, postlymphangitic, and postoperative lymphedema of the upper and lower extremities (Brunner, 1963; Herrmann, 1965; Gray et al., 1966; Krückemeyer and Scholz, 1967). These are, however, not considered here.

In a survey of the literature we found 115 cases of Stewart-Treves syndrome in the initial sense (Table 23). The patients, 113 women and 2 men, were aged 42–84 years at the time of diagnosis (average age: 63 years). Latent periods of the lymphangiosarcomas ranged from 1 to 26 years (average latent period: 10 years). Survival times were given in 89 cases: 42 patients died within the first year after the time of diagnosis, 31 died between the first and fifth year, and two died after survival times of more than 5 years. At the time of this publication 14 patients were alive; 5 of them had survival times of more than 5 years. On an average, the lymphangiosarcomas led to death after 16 months.

In most cases the tumors occurred on the interior surface of the lymphedematous upper arm. Initially small, hard, primary cutaneous efflorescences were observed. These prominences, which were bluish red to bluish black and had macular, papular, or nodular forms, grew rapidly and soon were surrounded by satellites, which were occasionally confluent. This process spread to the entire extremity and finally passed to the adjoining thoracic wall. Histologic examination of the tumors revealed angioplastic sarcomas with angiomatous or fibroblastic parts that were polymorphic and rich in mitoses; erythrocytic extravasations and hemosiderin deposits were present (Teller and Krüger, 1968). Metastases in the lungs, paraaortal lymph nodes, liver, and spleen were formed early. Fry et al. (1959) observed lung metastases in 60% of his patients.

Whereas there is general agreement about the angioplastic character of the tumors, many investigators doubted that all tumors originated from the lymph vessels. Jessner et al. (1952), McConnel and Haslam (1959), Wolff (1963), and Kleemann and Stiehl (1972) pointed out that the capillary tubes might also be involved in the formation of these tumors, and therefore used the term "angiosarcoma" instead of "lymphangiosarcoma."

There are various theories on the origin of these tumors:

1. Whereas most investigators considered the Stewart-Treves syndrome as an independent clinical picture, Laffargue et al. (1960), Martin and Vilain (1960), Ginnardi et al. (1960), and Salm (1963) interpreted the neoplastic changes as a special type of formation of metastases.
2. Stewart and Treves (1948) thought that a "systemic carcinogenic factor" had furthered the development of multiple tumors in their patients. Third primary tumors, however, occurred in only four of the other cases (one bronchial carcinoma, one carcinoma of the rectum, one skin carcinoma; Herrmann and Gruhn, 1957; Liszauer and Ross, 1957; Bachulis et al., 1967; Rytter, 1972).
3. Lymphedema, which was present for an average of 9 years, certainly played a decisive role in the formation of the tumors. Treves (1957), Britton and Nelson (1962), and Kappey (1967) found lymphedema in 41%, 13–57%, and 16% of their patients after breast amputation, respectively. Only in two cases did lymphedema and angiosarcomas occur simultaneously (Herrmann and Gruhn, 1957; Liszauer and Ross, 1957).
   Lymphedema can result from either an interruption of the lymph passage following the removal of the lymph nodes, or an occlusion of the lymph vessels (Wolff, 1963), which might be caused by chronic inflammation or pre- and postoperative irradiation (Treves, 1957; Wolff, 1963; Krückemayer and Scholz, 1967; Mackh, 1967). According to Hughes and Patel (1966) a venous stasis, hindering sufficient formation of collaterals, might be involved in the development of postoperative lymphedema of the upper arm.
   The pathogenesis of these angiosarcomas is thus far unclear. Wolff (1963) supposed that degenerative lesions of the vascular connective tissue constitute a tumor-favoring condition. McConnel and Haslam (1959), on the basis of histologic examinations, divided the development of sarcomas into three stages: In the first stage, the

collagenous fibers and the subcutaneous fat tissue in the edematous extremity degenerate. In the second stage, a premalignant angiomatosis is formed, which, in the third stage, changes to malignant sarcoma.

4. Most investigators excluded irradiation as a direct cause of the angiosarcomas since not all patients were irradiated, and tumors were also formed outside the irradiated areas (McConnel and Haslam, 1959; Herrmann, 1965; Gray et al., 1966; Kleeman and Stiehl, 1972; Rytter, 1972). In the present study ninty-five patients were irradiated, eleven were not irradiated, and in eight cases no statement was given.

McConnel and Haslam (1959), in a study of 894 mastectomized women, found Stewart-Treves syndrome in 0.45%. According to these authors the incidence of Stewart-Treves syndrome may be increased even to 10% among patients with severe chronic lymphedema. Angiosarcomas were observed not only after mastectomy, but also in primary or secondary lymphedema of other origin (Danese et al., 1967; Eby et al., 1967). The basic precondition for the development of these sarcomas is therefore the chronic edema. Becker (1970) and Campbell interpreted the Stewart-Treves syndrome as an iatrogenic disease, in which the mastectomy, by causing chronic lymphedema, provides the basis for the development of a second, more malignant tumor.

Finally, we would like to emphasize the difficulties in the differential diagnosis of the Stewart-Treves syndrome and the sarcoma haemorrhagicum idiopathicum multiplex Kaposi. Both the macroscopic and microscopic appearances of these tumors are very similar.

References

Andrews, G. C., Machanek, G. F. (1958): cited by Wolff, K. (1963)

Bachulis, B. L., Old, J. W., James, A. G.: Postmastectomy lymphangiosarcoma in a patient with carcinoma of the rectum. Amer. J. Surg. *113*, 289 (1967)

Barnett, W. O., Hardy, J. D., Hendrix, J. H.: Lymphangiosarcoma following post-mastectomy lymphedema. Ann. Surg. *169*, 960 (1969)

Becker, H.: Klinik und Pathologie der multiplen Malignome (163 Fälle). Med. Klin. *65*, 1775 (1970)

Birge, R. F., Peisen, C. J., Thornton, F. E., Powell, L. D. (1957): cited by Wolff, K. (1963)

Block, M. A., Fleming, J. L., Gish, J. R. (1956): cited by Wolff, K. (1963)

Boss, J. H., Urka, J.: Stewart-Treves syndrome. Angiosarcoma in postmastectomy lymphedema associated with disseminated fibrinoid vascular lesions. Amer. J. Surg. *101*, 248 (1961)

Bowers, W. F., Shaer, E. W., Legolvan, P. C.: Lymphangiosarcoma in the post-mastectomy lymphedematous arm. Amer. J. Surg. *90*, 682 (1955)

Brieler, H. S., Müller-Wiefel, H.: Stewart-Treves-Syndrom. Muench. med. Wschr. *115*, 1739 (1973)

Britton, N., Nelson, P. A.: Causes and treatment of postmastectomy lymphedema of the arm. J. Amer. med. Ass. *180*, 95 (1962)

Brunner, U.: Über das angioplastische Sarkom bei chronischem Lymphödem (Stewart-Treves-Syndrom). Schweiz. med. Wschr. *93*, 949 (1963)

Campbell: cited by Brunner, U. (1963)

Conte, A. J., Rella, A. J.: Angiosarcoma in lymphedema following mastectomy. N. Y. St. J. Med. *62*, 3966 (1972)

Cruse, R., Fisher, W. C., Usher, F. C.: Lymphangiosarcoma in postmastectomy lymphedema. Surgery *30*, 565 (1951)

Danese, C. A., Grishman, E., Dreiling, D. A.: Malignant vascular tumors of the lymphedematous extremity. Ann. Surg. *166*, 245 (1967)

Dembrow, V. D., Adair, F. E.: Lymphosarcoma in the postmastectomy lymphedematous arm. Cancer (Philad.) *14*, 210 (1961)

Di Simone, R. N., El-Mahdi, A. M., Hazra, T., Lott, S.: The response of Stewart-Treves syndrome to radiotherapy. Radiology *97*, 121 (1970)

Doremus, W. P., Salvia, G. A.: Lymphangiosarcoma in the postmastectomy lymphedematous arm. Amer. J. Surg. *96*, 576 (1958)

Duperrat, B., Jourdain, J., Noury, G., Laine, J.-Y.: Syndrome de Stewart-Treves. Bull. Soc. Franç. Derm. Syph. *78*, 650 (1971)

Eby, Ch. S., Brennan, M. J., Fine, G.: Lymphangiosarcoma: A lethal complication of chronic lymphedema. Arch. Surg. (Chicago) *94*, 223 (1967)

Ende, M.: Lymphangiosarcoma. Pacif. Med. Surg. *74*, 80 (1966)

Ferraro, L. R.: Lymphangiosaroma in postmastectomy lymphedema. Cancer *3*, 511 (1950)

Fisher, J. H.: Postmastectomy lymphangiosarcoma in the lymphedematous arm. Canad. J. Surg. *8*, 350 (1965)

Fitzpatrick, P. J.: Lymphangiosarcoma and breast cancer. Canad. J. Surg. *12*, 172 (1969)

Froio, G. F., Kirkland, W.: Lymphangiosarcoma in postmastectomy lymphedema. Ann. Surg. *135*, 421 (1952)

Fry, W. J., Campbell, D. A., Coller, F. A. (1959): cited by Wolff, K. (1963)

Ginnardi, G. F., Pelie, G., Zampi, G.: Le quadro clinico e la base istopatologische della cosidetta sindrome di Stewart e Treves. Arch. Vecchi Anat. Path. *34*, 361 (1960)

Gray, G. F. Jr., Gonzales-Licea, A., Hartmann, W. H., Woods, A. C.: Angiosarcoma in lymphedema, an unusual case of Stewart-Treves syndrome. Bull. Johns Hopk. Hosp. *119*, 117 (1966)

Hall-Smith, S. P. H., Haber, H.: Lymphangiosarcoma in postmastectomy lymphoedema (Stewart-Treves syndrome). Proc. roy. Soc. Med. *47*, 174 (1954)

Herrmann, J. B.: Lymphangiosarcoma of the chronically edematous extremity. Surg. Gynec. Obstet. *121*, 1107 (1965)

Herrmann, J. B., Ariel, I. M.: Therapy of lymphangiosarcoma of the chronically edematous limb. Five years cure of a patient by intra-arterial radioactive yttrium. Amer. J. Roentgenol. *99*, 393 (1967)

errmann, J. B., Gruhn, J. G. (1957): cited by Wolff, K. (1963)

Hilfinger, M. F., Eberle, R. D.: Lymphangiosarcoma in postmastectomy lymphedema. Cancer (Philad.) *6*, 1192 (1953)

Hope-Stone, H. F., Bence, E. A. (1954): cited by Wolff, K. (1963)

Hughes, J. H., Patel, A. R.: Swelling of the arm following radical mastectomy. Brit. J. Surg. *53*, 4 (1966)

Hume, H. A., Erb, W. H., Stevens, L. W.: Lymphangiosarcoma following radical mastectomy. Surg. Gynec. Obstet. *116*, 117 (1963)

Jansey, F., Syato, P. B., Wright, A. (1957): cited by Wolff, K. (1963)

Jessner, M., Zak, F. G., Rein, Ch. R.: Angiosarcoma in postmastectomy lymphedema (Stewart-Treves syndrome). Arch. Derm. *65*, 123 (1952)

Kappey, F.: Das angioplastische Sarkom bei chronischem Lymphödem nach Ablatio Mammae (Stewart-Trèves-Syndrom). Chirurg *38*, 59 (1967)

Kettle, J. H.: Lymphangiosarcoma following post-mastectomy lymphoedema. Brit. med. J. *1*, 193 (1957)

Kleemann, W., Stiehl, P.: Zur Ätiologie und Histologie des Stewart-Treves-Syndroms. Zbl. allg. Path. path. Anat. *116*, 147 (1972)

Krückemeyer, K., Scholz, H.: Über ein angioblastisches Sarkom bei chronischem Lymphödem nach Mamma-Radikaloperation (Stewart-Treves-Syndrom). Zbl. Gynaek. *89*, 229 (1967)

Laffargue, P., Pinet, F., Le Go, R.: Syndrome de Stewart et Treves (Métastases épithéliomateuses on angiosarcome dans les gros bras compliquant la mammaectomie). Presse méd. *68*, 1506 (1960)

Liszauer, S., Ross, R. C. (1957): cited by Wolff, K. (1963)

Mackh, G.: Das Stewart-Treves-Syndrom. Bruns Beitr. klin. Chir. *214*, 235 (1967)

Marshall, J. F.: Lymphangiosarcoma of the arm following radical mastectomy. Ann. Surg. *142*, 871 (1955)

Martin, E., Vilain, R.: Discussion anatomique sur un cas de syndrome de Stewart-Treves. Sem. Hôp. Paris *8*, 246 (1960)

McCarthy, W. D., Pack, G. T.: Malignant blood vessel tumors. Report of 56 cases of angiosarcoma and Kaposi's sarcoma. Surg. Gynec. Obstet. *91*, 465 (1950)

McConnel, E. M., Harris, H. R.: Angiosarcoma in postmastectomy lymphedema. Brit. J. Surg. *53*, 572 (1966)

McConnel, E. M., Haslam, P.: Angiosarcoma in postmastectomy lymphedema; a report of five cases and a review of the literature. Brit. J. Surg. *46*, 322 (1959)

McSwain, B., Stephenson, S.: Lymphangiosarcoma of the edematous extremity. Ann. Surg. *151*, 649 (1960)

Nelson, W. R., Morfit, H. M.: Lymphangiosarcoma in the lymphedematous arm after radical mastectomy. Cancer (Philad.) *9*, 1189 (1956)

Nemoto, T., Stubbe, N., Gaeta, J., Dao, T.: Pathogenesis of lymphangiosarcoma following mastectomy and irradiation. Surg. Gynec. Obstet. *128*, 489 (1969)

Oettle, A. G., van Blerk, P. J. P.: Postmastectomy lymphostatic endothelioma of Stewart and Treves in a male. Brit. J. Surg. *50*, 736 (1963)

Ogilvy, W. L., Franklin, R. H., Aird, I. (1959): cited by Wolff, K. (1963)

Oota, K., Baba, T.: A case of postmastectomy lymphangiosarcoma. Gann *147*, 748 (1956)

Patton, R. J. (1958): cited by Wolff, K. (1963)

Ramsey, H. E., Lucas, J. C., Gray, G.: Post-mastectomy angiosarcoma in the male. J. nat. med. Ass. *60*, 468 (1968)

Rawson, A. J., Frank, J. L.: Treatment by irradiation of lymphangiosarcoma in postmastectomy lymphedema. Cancer (Philad.) *6*, 269 (1953)

René, L., Bolgert, M., Le Sourd, M., Tabernat, J., Poisson, R.: Un cas de syndrome de Stewart-Treves. Bull. Soc. franç. Derm. Syph. *70*, 9 (1963)

Riddel, R. J.: Lymphangioendothelioma of the arm following radical mastectomy for carcinoma of the breast. Aust. J. Surg. *30*, 229 (1960)

Rytter, M.: Metachrones angioplatisches Sarkom (Stewart-Treves-Syndrom) mit synchronem Basalzellkarzinom in einem chronisch lymphgestauten Arm nach Ablatio mammae. Dtsch. Gesundh.-Wes. *27*, 118 (1972)

Rytter, M., Nitzschner, H.: Das Stewart-Treves-Syndrom. Med. Bild *13*, 179 (1970)

Salm, R.: The nature of the so-called postmastectomy lymphangiosarcoma. J. Path. Bact. *85*, 445 (1963)

Silverberg, S. T., Kay, S., Koss, L. G.: Postmastectomy lymphangiosarcoma: Ultra-structural observations. Cancer (Philad.) *27*, 100 (1971)

Southwick, H. W., Slaughter, D. P.: Lymphangiosarcoma in postmastectomy lymphedema. Cancer (Philad.) *8*, 158 (1955)

Staff: Stewart-Treves syndrome (postmastectomy lymphangiosarcoma). Arch. Derm. *102*, 341 (1970)

Sternby, N. H., Gynning, I., Hogemann, K. E. (1961): cited by Wolff, K. (1963)

Stewart, F. W., Treves, N.: Lymphangiosarcoma in postmastectomy lymphedema: A report of six cases in elephantiasis chirurgica. Cancer (Philad.) *1*, 64 (1948)

Taswell, H. F., Soule, E. H., Coventry, M. B.: Lymphangiosarcoma arising in chronic lymphedematous extremities. Report of thirteen cases and review of literature. J. Bone Jt. Surg. *44*, 272 (1962)

Teller, H., Krüger, H.: Gegenüberstellung von Angiomatosis Kaposi (Sarcoma idiopathicum haemorrhagicum multiplex) und Stewart-Treves-Syndrom (Sarcoma angioplasticum in elephantiasi) XIII Congr. Internat. Dermatol., 1967. Berlin-Heidelberg-New York: Springer 1968

Tentschov, G., Andrev, V., Raitschev, R., Kristev, B. (1961): cited by Wolff, K. (1963)

Toujas, L., Ferrand, B., Guelfi, J., Illes, J.: Syndrome de Stewart-Treves. Etudes ultrastructurales d'un cas. Bull. Ass. Anat. (Nancy) *139*, 1150 (1968)

Treves, N.: An evaluation of the etiological factors of lymphedema following radical mastectomy. Cancer (Philad.) *10*, 444 (1957)

Vos, P. A.: Lymphangiosarcoma in post-mastectomy lymphoedema. Arch. Chir. Neerl. *4*, 197 (1952)

Wilson, R.: Lymphangiosarcoma in the postmastectomy lymphedematous arm. Canad. J. Surg. *5*, 208 (1962)

Wolff, K.: Das Stewart-Treves-Syndrom. Arch. klin. exp. Derm. *216*, 468 (1963)

Table 23. Stewart-Treves syndrome

Patient Age	Sex	Interval between mastectomy and angiosarcoma (years)	Reference
52	F	9	Stewart and Treves, 1948
60	F	8	„
69	F	10	„
46	F	9	„
64	F	24	„
56	F	6	„
66	F	9	Ferraro, 1950
67	F	10	McCarthy and Pack, 1950
69	F	10	„
65	F	12	Cruse et al., 1951
54	F	11	Froio and Kirkland, 1952
61	F	7	Jessner et al., 1952
55	F	14	Vos, 1952
52	F	10	Hilfinger and Eberle, 1953

Table 23 (continued)

| Patient | | Interval between mastectomy | Reference |
Age	Sex	and angiosarcoma (years)	
63	F	16	Hilfinger and Eberle, 1953
70	F	14	Rawson and Frank, 1953
72	F	2	Hall-Smith and Haber, 1954
47	F	6	Bowers et al., 1955
71	F	5	Marshall, 1955
49	F	8	Southwick and Slaughter, 1955
77	F	9	Nelson and Morfit, 1956
71	F	9	,,
–	F	8	Block et al., 1956
69	F	5	Oota and Baba, 1956
72	F	8	Kettle, 1957
55	F	2	Birge et al., 1957
59	F	13	Herrmann and Gruhn, 1957
75	F	9	,,
68	F	14	,,
66	F	10	,,
44	F	8	,,
65	F	9	Jansey et al., 1957
62	F	4.5	Liszauer and Ross, 1957
73	F	14.5	Andrews and Machanek, 1958
72	F	11	Doremus and Salvia, 1958
75	F	17	Patton, 1958
48	F	5	Fry et al., 1959
73	F	23	,,
46	F	11	Hope-Stone and Bence, 1959
76	F	7	,,
52	F	11	McConnel and Haslam, 1959
52	F	8	,,
58	F	8	,,
56	F	7	Ogilvy et al., 1959
71	F	14	Ginnardi et al., 1960
60	F	8	Laffargue et al., 1960
57	F	10	Martin and Villain, 1960
64	F	11	McSwain and Stephenson, 1960
84	F	23	Riddel, 1960
54	F	6	,,

Table 23 (continued)

Patient Age	Sex	Interval between mastectomy and angiosarcoma (years)	Reference
54	F	7.5	Boss and Urka, 1961
54	F	12	Dembrow and Adair, 1961
69	F	1.5	Sternby et al., 1961
59	F	9.5	,,
55	F	11	Tentschov et al., 1961
68	F	8	Conte and Rella, 1962
69	F	5	Taswell et al., 1962
54	F	4.5	,,
47	F	8.5	,,
55	F	14	,,
50	F	9	,,
58	F	11	,,
69	F	5	,,
64	F	12	,,
66	F	7	,,
67	F	7	,,
57	F	10	,,
50	F	10	Wilson, 1962
60	F	8	Hume et al., 1963
65	F	8	,,
71	M	8	Oettle and van Blerk, 1963
61	F	4	René et al., 1963
69	F	26	Wolff, 1963
75	F	8	Fisher, 1965
60	F	5	,,
42	F	11	,,
68	F	10	,,
48	F	8	Ende, 1966
55	F	10	McConnel and Harris, 1966
60	F	12	,,
78	F	8	,,
55	F	6	,,
74	F	5	Bachulis et al., 1967
68	F	18	Eby et al., 1967
68	F	7	,,
63	F	21	Herrmann and Ariel, 1967

Table 23 (continued)

Patient Age	Sex	Interval between mastectomy and angiosarcoma (years)	Reference
69	F	16	Kappey, 1967
74	F	10	Krückemeyer and Scholz, 1967
75	F	10	Mackh, 1967
72	F	8	,,
57	M	9	Ramsey et al., 1968
73	F	8	Toujas et al., 1968
78	F	11	Fitzpatrick, 1969
61	F	8	,,
72	F	8	,,
76	F	9	,,
66	F	10	,,
73	F	15	,,
64	F	9	,,
67	F	13	Barnett et al., 1969
52	F	7	,,
56	F	15	,,
57	F	7	Nemoto et al., 1969
74	F	9	,,
73	F	8	,,
66	F	9	,,
47	F	2	Di Simone et al., 1970
64	F	7.5	,,
67	F	15	,,
61	F	12	Rytter and Nitzschner, 1970
80	F	16	Staff, 1970
82	F	10	Duperrat et al., 1971
56	F	7	Brieler and Müller-Wiefel, 1973
63	F	7	Silverberg et al., 1971
51	F	12	,,

Tumors of the Colon Following Ureterosigmoidostomy

Ureterosigmoidostomy is used as a method for eliminating the urine after radical cystectomy, vesical ectopia, incontinence, and irreversible ureteral stricture. Complications of this method include chronic pyelonephritis and hyperchloremic acidosis due to reabsorption of urinary electrolytes from the intestine.

97

Hammer (1929) reported the first case of tumor following this method. The tumor originated in a bladder segment and then passed to the colon. Dixon and Weisman (1948) reported a further case, in which ureterosigmoidostomy was followed by a primary tumor of the colon.

In a survey of the literature we found 25 cases of colonic tumors associated with ureterosigmoidostomy; these are shown in Table 24. The patients (18 men and seven women) were aged 17–82 years at the time of diagnosis (average age: 40 years). Indications for the ureterosigmoidostomy were vesical ectopia (13 cases), abacterial cystitis (three cases), epispadia (two cases), congenital incontinence, persisting urinary fistula, and papilloma and carcinoma of the bladder (one case each). Latent periods of the tumors ranged from 7 to 46 years (average latent period: 22 years). Tumor localizations were described in 16 cases: left anastomosis (nine cases), right anastomosis (five cases), both anastomoses (two cases). Histologic examination revealed 16 adenocarcinomas; in one patient an adenocarcinoma and an aplastic carcinoma occurred simultaneously. In four cases the tumors were histologically classified as "carcinoma"; four patients had benign tumors that were histologically diagnosed as adenomatous polyps. In 15 patients the course of the disease could be traced: eight patients died with metastases, five of them within the first year after the diagnosis; seven patients, five of whom had survived for more than 1 year, were without relapse at the time of this publication.

The following theories have been considered to explain the pathogenesis of these tumors:

1. Chronic inflammatory influence of the urine (Kozak, 1966) or mechanic traumatic effects of the operation (Gillman, 1964) led to neoplastic alterations in the colon mucosa.
2. Persisting mechanic irritation of the colon mucosa by the ureter led to mucosal ulceration and reconstruction with the formation of epithelial hyperplasias (Gillman, 1964).
3. Formation of tumor on the basis of pseudopolyps: The lower end of the ureter, projecting into the colon, includes a pseudopolyp that is more solid than the surrounding parts of the colon. This pseudopolyp is therefore exposed to traumatic influences from the fecal stream.
4. Formation of tumor by carcinogenic ingredients of the urine (Gillman, 1964). Scott and Boyd (1953) induced tumors of the colon after ureterosigmoidostomy in dogs and subsequent application of β-naphthylamine, the urinary metabolites of which are known bladder carcinogens. This theory, however, does not take into consideration those cases of tumor in which nephrectomy was used and the uretal stump retained.

Twenty-four out of 100,000 individuals of age 45, according to the US National Health Statistics, develop tumors of the colon, whereas in patients with ureterosigmoidostomy this number is expected to increase to 13,300.

In the present cases the average age (40 years) was significantly below that observed in other tumors of the colon (50–70 years). Only three patients were older than 60 years; 16 patients were younger than 45 years. The average interval between the ureterosigmoidostomy and the tumor diagnosis was 20 years.

References

Aldis, A. S.: Carcinoma of the colon following transplantation of the ureter, and at the site of transplantation. Proc. Roy. Soc. Med. *54*, 159 (1961)

Brekkan, E., Colleen, S. Myrvold, H., Du Rietz, B., Schnürer, L. B., Fritjofsson, A.: Colonic neoplasia: A late complication of ureterosigmoidostomy. Scand. J. Urol. Nephrol. *6*, 197 (1972)

Dixon, C. F., Weismann, R. E.: Polyps of sigmoid occurring 30 years after bilateral ureterosigmoidostomy for exstrophy of bladder. Surgery *24*, 1026 (1948)

Ellis, F. G.: Adenocarcinoma complicating ureterosigmoidostomy. Proc. Roy. Soc. Med. *55*, 100 (1962)

Ferguson, L. K. (1966): cited by Urdaneta, L. F., Duffell, D., Greevy, C. D., Aust, J. B. (1966)

Gillman, J. C.: Adenomatous polyp of bowel following ureterocolic anastomosis. Brit. J. Urol. *36*, 264 (1964)

Hammer, E.: Cancer du colon sigmoide dix ans après implantation des uretèrs d'une vessie éxtrophié. J. Urol. Nephrol. *28*, 260 (1929)

Haney, M. J., McGarity, W. C.: Ureterosigmoidostomy and neoplasms of the colon. Arch. Surg. (Chicago) *103*, 69 (1971)

Kille, J. N., Glick, S.: Neoplasia complicating ureterosigmoidostomy. Brit. Med. J. *4*, 783 (1967)

Kozak, J. A., Watkins, W. E., Jewell, W. R.: Neoplastic stomach obstruction: A complication of ureterosigmoidostomy. J. Urol. *96*, 691 (1966)

Massachusetts General Hospital Case Records (Case No. 44052): New Engl. J. Med. *258*, 244 (1958)

Mueller, C. E., Thornbury, J. R.: Adenocarcinoma of the colon complicating ureterosigmoidostomy: A case report and review of the literature. J. Urol. *109*, 225 (1973)

Oetjen, L. H., Campbell, J. L., Thomley, M. W., Parsons, R. L.: Carcinoma of the colon following ureterosigmoidostomy: Report of a case. J. Urol. *104*, 536 (1970)

Richter, R. M., Ginsberg, S. A.: Late development of colonic carcinoma complicating ureterosigmoidostomy. Amer. J. Surg. *113*, 843 (1967)

Scheinman, L. J.: Tumor at site of ureterosigmoidostomy nine years postoperatively. J. Urol. *85*, 934 (1961)

Scott, W. W., Boyd, H. L.: A study of the carcinogenic effect of betanaphthylamine on the normal and substituted sigmoid loop ladder in dogs. J. Urol. *70*, 914 (1953)

Sugg, W. L.: Tumour at site of ureterosigmoidostomy: Report of a case and review of the literature. Ann. Surg. *155*, 572 (1962)

Urdaneta, L. F., Duffell, D., Greevy, C. D., Aust, J. B.: Late development of primary carcinoma of the colon following ureterosigmoidostomy: Report of three cases and literature review. Ann. Surg. *164*, 503 (1966)

Whitaker, R. H., Pugh, R. C. B., Dow, D.: Colonic tumours following ureterosigmoidostomy. Brit. J. Urol. *43*, 562 (1971)

Wilson, L. L.: Carcinomatous ureteric obstruction thirty years after ureterosigmoidostomy. Aust. J. Surg. *27*, 158 (1957/58)

Table 24. Colonic tumors following ureterosigmoidostomy

Patient Age	Sex	Type of colonic tumor	Reference
33	M	Adenomatous polyp	Dixon and Weisman, 1948
47	F	Anaplastic carcinoma	Wilson, 1957/58
63	M	Adenocarcinoma	Massachusetts General Hospital, 1958
33	F	Anaplastic carcinoma	Aldis, 1961
51	M	Adenocarcinoma	Scheinman, 1961
55	M	Adenomatous polyp	Ellis, 1962
21	F	Adenocarcinoma	Sugg, 1962
82	M	3 adenomatous polyps	Gillman, 1964
30	M	Polypoid adenocarcinoma	Ferguson, 1966
17	M	Adenocarcinoma	Kozak et al., 1966
40	F	Adenocarcinoma	Urdaneta et al., 1966
64	M	Adenocarcinoma	,,
34	F	Undifferentiated carcinoma	,,
26	M	Adenocarcinoma	Kille and Glick, 1967
19	M	Adenomatous polyp	,,
24	M	Adenocarcinoma	Richter and Ginsberg, 1967
23	M	Adenocarcinoma	Oetjen et al., 1970
57	M	Adenocarcinoma	Whitaker et al., 1971
22	M	Adenocarcinoma	,,
39	M	Anaplastic carcinoma	,,
44	F	Adenomatous polyp	Haney and McGarity, 1971
23	F	Anaplastic carcinoma	Brekkan et al., 1972
–	–	Adenocarcinoma	,,
44	M	Adenocarcinoma	,,
47	M	Adenocarcinoma	,,
55	M	Adenocarcinoma	Mueller and Thornbury, 1973

Appendix

Drugs that either showed carcinogenic effects only in experimental studies or to which a few questionable cases of human tumor were attributed are considered in this chapter.

Isonicotinic acid hydrazide (INH), which is used in the chemotherapy of tuberculosis, induced lung adenomas in mice (Juhasz et al., 1957; Mori et al., 1960; Biancifiori and Severi, 1966), whereas it was noncarcinogenic in hamsters (Toth and Boreisha, 1969) and rats (Peacock and Peacock, 1966). Hammond et al. (1967), in a study of 311 tuberculosis patients treated with INH during the years 1951–1956, 502 women treated with INH during pregnancy, and 660 children intrauterinely exposed to INH, found no increase in the tumor incidence of these patients. In a study of 150 patients with skin diseases Pompe (1956) found that the incidence of carcinomas arising from lupus increased from 0.5% to 4.6% after treatment with INH. However, later the author doubted his statistical findings and attributed this increase to other factors. Michalowsky and Kudejko (1965), Nyfors (1968), and Jung (1971), who also investigated this question, found no indications of an increased risk of cancer in INH-treated patients with dermatoses.

Griseofulvin, a fungal antibiotic, induced tumors of the liver in newborn mice after subcutaneous application of high doses (Epstein et al., 1967). Barich et al. (1962) found "co-carcinogenic" effects of griseofulvin on the skin of mice. Götz and Reichenberger (1972), in an inquiry among 1670 German dermatologists, found no indications of carcinogenic effects of Grisefulvin on the skin of man.

Tannin was used, especially during the war, for the local treatment of burns. It showed toxic effects upon the liver of animals and man (for references see Korpassy, 1961). In rats it induced tumors of the liver after subcutaneous application (Korpassy, 1961), whereas in mice it was noncarcinogenic (Bichel and Bach, 1968). From human pathology we have thus far no indications of liver carcinogenic effects of this substance. Carcinogenic effects after oral administration of tannin are not expected since it is not absorbed by the gastrointestinal tract.

Thiouracil, which is used in the treatment of thyropathies, and its derivatives induced malignant, transplantable tumors of the thyroid gland in animal experiments (Purves et al., 1951; Sellers, 1953).

Tumors of the thyroid gland in man following medication with thiouracil were observed by Payne et al. (1947), Dufour and Cabanie (1948), Pemberton and Black (1948), Herrmann (1951), Minder (1952), Boulet and Barjon (1953), and Amrhein et al. (1970). Lundsgaard-Hansen (1956), Reveno (1964), and Hershman et al. (1966), however, found no increased incidence of these tumors in large groups of thiouracil-treated patients.

Glucocorticoids did not induce tumors in animal studies (Schmähl and Habs, 1976); they did, however, increase the rate of metastases of transplantable tumors, probably by means of their immunodepressive effects (Agosin et al., 1952; Albert and Zeidman, 1962). In autochthonous animal tumors the formation of metastases was not favored by these substances (Schmähl, 1974, 1975; Schmähl and Habs, 1976).

Some years ago oncogenic SV 40 viruses were detected experimentally (Girardi et al., 1962) in *poliomyelitis vaccine*. SV 40-induced tumors in man, however, were not reported.

Cyclamate, which is used as an artificial sweetener in the treatment of diabetes and in diet cures, was accused of having carcinogenic effects on the bladder of experimental animals (Price et al., 1970; Editorial, Lancet, 1970). This led to a ban of cyclamate in the United States and in some European countries. In the meantime the findings of Price et al. were contradicted by numerous experimental studies (Schmähl and Krüger, 1972; Schmähl, 1973; for further references see Althoff et al., 1975). Tumors in man following the use of cyclamate have thus far not been observed.

For years possible carcinogenic effects of the long-term use of *ovulation inhibitors* have been the subject of discussion. Statistical studies showed that the general risk of cancer was not increased in women treated with these drugs; in particular, an increase of carcinomas of the portio, endometrium, and breast was not observed (for references see Hillemanns, 1971). Ovulation inhibitors habe not yet been shown in experimental studies to have carcinogenic effects (Thomas et al., 1972), except in one (adequate?) study in which benign and malignant tumors of the breast occurred in beagle dogs after 7 years' treatment with megastrol acetate. Therefore various contraceptives containing this compound were withdrawn from the market (Editorial, Deutsches Ärzteblatt, 1976).

Baum et al. (1973) and Contostavlos (1973) observed hepatomas in younger women who had taken ovulation inhibitors for several years. These observations were confirmed by other investigators (Berg et al., 1974; Frederick et al., 1974; Horvarth et al., 1974; Kelso, 1974; Knapp and Rueber, 1974; Mays et al., 1974; Meyer et al., 1974; O'Sullivan and Wilding, 1974; Tauntas et al., 1974; Amariks and Thompson, 1975; Antionades and Brooks, 1975; Stenwig and Solgaard, 1975; Edmondson et al., 1976). The last authors attributed the liver toxicity to the mestranol contained in these drugs. Up to now almost 100 cases of hepatoma have been reported, which had the following characteristics: All women were aged between 30 and 40 years at the time of the diagnosis, and had taken ovulation inhibitors for about 5 years. They developed liver tumors, most of which were histologically benign hepatomas. Less frequently liver cell carcinomas (Hermann and David, 1973; Berg et al., 1974; Meyer et al., 1974), focal, nodulary hyperplasias (Mays et al., 1974), and hamartomas (O'Sullivan and Wilding, 1974) were formed. In 15 cases the tumors were diagnosed after rupture and hemorrhages into the abdomen. In the nontumorous hepatic tissue, thrombus formation as well as signs of severe blood stases, which were sometimes along the lines of a peliosis hepatis or a Budd-Chiari syndrome (Fox and Lahcen, 1974; Hoyumpa et al., 1971), were occasionally observed, but no cirrhotic changes were noted. Electronmicroscopic examinations revealed abnormal mitochondria (polymorphic giant mitochondria), which were also present in the livers of healthy women taking ovulation inhibitors (Horvarth et al., 1974).

Although an exact causal relationship between treatment with ovulation inhibitors and tumors of the liver is so far not established, the frequent recent observations support this possibility. Further support in this respect is given by the findings of Horvarth et al. (1974) who described hepatomas in rats after application of high doses of northisterone and megestrol. Ovulation inhibitors have been in common use for about two decades. It is therefore remarkable that observations of hepatomas in women taking these drugs have accumulated only during the last few years, although the induction period of this tumor is rated at only 5 to 6 years.

Hepatomas and liver changes along the lines of a peliosis hepatis were also observed after treatment with various *androgenic and anabolic steroids* (Johnson et al., 1972; Naeim et al., 1973; Meadows et al., 1974; Bagheri and Boyer, 1974; Farrel et al., 1975). Observations of Donald and Smith (1975) and Ziel and Finkle (1975), who found endometrial carcinomas in women treated with estronsulfate-containing contraceptives, were thus far not confirmed.

It has been suspected for a long time that estrogens, the carcinogenic action of which was proved experimentally, may have carcinogenic effects in man. Recent communications give, indeed, rise to the suspicion that treatment with high doses of estrogens carries an increased risk of cancer of the breast and the endometrium (Gordan and Greenberg, 1976; Hoover et al., 1976). Since estrogens have a considerable therapeutic benefit in the treatment of many diseases but, on the other hand, seem to be of a minor carcinogenic risk, one should consider more careful their range of indications in the future and should use them only for established indications. In addition, women receiving estrogens should be observed carefully.

Recent epidemiologic findings indicated a possible role of *Rauwolfia derivatives (Reserpine)* in the initiation of human breast cancer (Boston Collaborative Drug Program, 1974; Heinonen et al., 1974; Armstrong et al., 1974). These findings were, however, contradicted by Mack et al. (1975), Laska et al. (1975), and O'Fallon et al. (1975). We believe that the use of this drug, because of its important therapeutic effect, should not be discontinued (see also the final statement of the Rauwolfia-Committee of the Federal Board of Health on the question of an increased risk of breast cancer by Rauwolfia preparations, Deutsches Ärzteblatt, 1976).

Spironolactones were also suspected of inducing breast cancer in man (Loube and Ruirk, 1975). Zick and Armstrong (1975), however, contradicted this supposition.

Experimental findings by Schmähl and Steinhoff (1973) indicate that *alkylating triazene derivatives and nitrosoureas,* which are occasionally used in cancer chemotherapy or immunosuppression, are, because of their alkylating properties, carcinogenic in man also.

The same applies to *streptozotocin* which is occasionally used in the therapy of pancreatic carcinomas (for references see DuPriest et al., 1975). This substance induced tumors of the pancreas in experimental studies (for references see Rudas, 1972, and Mauer et al., 1974).

Recent experimental findings of Toth (1976), which have not yet been reproduced, indicated carcinogenic effects of *hydrazine derivatives.* β-*Phenylethylhydrazine sulfate,* which is used as an antidepressant, induced malignant tumors in the lungs and blood vessels of mice.

Extrapolation of experimental findings with the following substances to human conditions is particularly difficult since these drugs either were carcinogenic in only one animal species, or the experimental findings could not be reproduced.

Hexamethylenetetramine (urotropin) induced local sarcomas in rats (Watanabe and Sugimoto, 1955), a finding that could not be confirmed by Della Porta et al. (1968).

The suspicion that *herbae artemesiae absinthii* has carcinogenic effects could not be verified by Schmähl (1956).

Pronethalol (alderlin), metronidazole (clont), and niridazole (ambilhar) exhibited carcinogenic effects in mice (Howe, 1965; Rustia and Shubik, 1972; Urman et al., 1975). The findings with the latter substance are remarkable in that some *nitrofurane derivatives,* which are used predominantly in the treatment of urologic infections, were potent carcinogens in animal experiments (Price et al., 1966).

Experimental findings with *cantharidin and asiaticosides* (Laerum and Iversen, 1972), which induced local skin tumors and reticuloses in mice, are hardly transferable to human conditions. The same applies to clinical observations of isolated cases of tumor following the use a drug. as was the case with the observation of Garson et al. (1969) who found acute leukemia following treatment with *lysergide.*

Recent experimental findings of Corbett (1976) on potential carcinogenic effects of *inhalation anesthetics* need further experimental confirmation before they can be discussed further.

Drugs that have merely a historic significance are not considered here. They are, however, for the sake of completeness included in Table 25, in which the drugs mentioned here are classified as certain/probable, possible, and unprobable/not assessable carcinogens.

Experimental findings that, in the stomach of experimental animals and probably in the human stomach, too, highly carcinogenic nitrosamines can be formed from some *drugs containing* secondary (or tertiary) *amines* [i. e., *aminopyrine (Pyramidon)*] and nitrite, which occurs naturally in various foodstuffs and also in human saliva (Lijinsky et al., 1972; Lijinsky and Greenblatt, 1972; Lijinsky and Singer, 1974) may prove to be of great practical importance: Two per se noncarcinogenic substances, one of which may be a drug, may lead to tumors induced by endogenously formed nitrosamines. Tables 26 and 27 show drugs that were found to act as precursors of nitroso-compounds. The carcinogenicity of several of these compounds has been established in almost 20 animal species.

References

Agosin, M., Christen, R., Badinez, O., Gasic, G., Neghme, A., Pizarro, O., Jarpa, A.: Cortisone-induced metastases of adenocarcinoma in mice. Proc. Soc. Exp. Biol. (N. Y.) *80,* 128 (1952)

Albert, D., Zeidman, I.: Relation of glucocorticoid activity of steroids to number of metastases. Cancer Res. *22,* 1297 (1962)

Althoff, J., Cardesa, A., Pour, P., Shubik, P.: A chronic study of artificial sweeteners in Syrian golden hamsters. Cancer Letters *1,* 21 (1975)

Amariks, J. A., Thompson, N. W.: Hepatic cell adenomas, spontaneous liver rupture and oral contraceptives. Arch. Surg. (Chicago) *110,* 548 (1975)

Amrhein, J. A., Kenny, F. M., Ross, D.: Granulocytopenia, lupus-like syndrome, and other complications of propylthiouracil therapy. J. Pediat. *76*, 54 (1970)

Antionades, K., Brooks, C. E.: Hemioperitoneum from liver cell adenomas in a patient on oral contraceptives. Surgery *77*, 137 (1975)

Armstrong, B., Stevens, N., Toll, R.: Retrospective study of the association between use of rauwolfia derivatives and breast cancer in English women. Lancet *2*, 672 (1974)

Bachmann, W. E., Deno, N. C.: The nitrosation of hexamethylenetetramine and related compounds. J. Amer. Chem. Soc. *73*, 2777 (1951)

Bagheri, S. A., Boyer, J. L.: Peliosis hepatis associated with androgenic-anabolic steroid therapy. A severe form of hepatic injury. Ann. Intern. Med. *81*, 610 (1974)

Barich, L. L., Schwarz, J., Barich, D.: Oral griseofulvin: A cocarcinogenic agent to methylcholanthrene-induced cutaneous tumors. Cancer Res. *22*, 53 (1962)

Baum, J. K., Holtz, F., Bookstein, J. J., Klein, E. W.: Possible association between benign hepatomas and oral contraceptives. Lancet *2*, 926 (1973)

Berg, J. W., Ketelaar, R. J., Rose, E. F., Vernon, R. G.: Hepatomas and oral contraceptives. Lancet *2*, 349 (1974)

Biancifiori, C., Severi, L.: The relation of isoniacid and allied compounds to carcinogenesis in some species of small laboratory animals: A review. Brit. J. Cancer *20*, 528 (1966)

Bichel, J., Bach, A.: Investigations on the toxicity of small chronic doses of tannic acid with special reference to possible carcinogenicity. Acta Pharmacol. *26*, 41 (1968)

Boston Collaborative Drug Program: Reserpine and breast cancer. Lancet *2*, 669 (1974)

Boulet, P., Barjon, P.: Cancer thyroidien découvert chez deux basedowiennes apparent guéries par antithyroidiens. Ann. Endocr. *14*, 927 (1953)

Contostavlos, D. L.: Benign hepatomas and oral contraceptives. Lancet *2*, 1200 (1973)

Corbett, T. H.: Cancer and congenital anomalies associated with anesthetics. Ann. N.Y. Acad. Sci. *271*, 58–66 (1976)

Della Porta, G., Colnaghi, M. J., Parmiani, G.: Non-carcinogenicity of hexamethylenetetramine in mice and rats. Food Cosmet. Toxicol. *6*, 707 (1968)

Donald, C., Smith, C. S.: Association of exogenous estrogen and endometrial carcinoma. New Engl. J. Med. *293*, 2264 (1975)

Dufour and Cabanie (1948): cited by Lundsgaard-Hansen, P. (1956)

DuPriest, R. W., Huntington, M. C., Massey, W. H.: Streptozotocin therapy in 22 cancer patients. Cancer (Philad.) *35*, 358 (1975)

Editorial: Zur Risikoabwägung bei der Verschreibung von weiblichen Sexualhormonen. Dtsch. Ärzteblatt *73*, 3 (1976)

Editorial: Why cyclamates were banned. Lancet *1*, 1091 (1970)

Edmondson, H. A., Henderson, B., Benton, B.: Liver cell adenomas associated with use of oral contraceptives. New Engl. J. Med. *294*, 470 (1976)

Eisenbrand, G., Preussmann, R.: Nitrosation of Phenacetin. Formation of N-nitroso-2-nitro-4-ethoxyacetanilide as an unstable product of nitrosation in dilute aqueous-acid solution. Arzneimittel-Forsch. *25*, 1472 (1975)

Epstein, S. S., Andrea, J., Joshi, S., Mantel, N.: Hepatocarcinogenicity of griseofulvin following parenteral administration to infant mice. Cancer Res. *27*, 1900 (1967)

Farrel, G. C., Joshua, D. E., Uren, R. F., Baird, P. J.: Androgen-induced hepatoma. Lancet *1*, 430 (1975)

Fox, K. A., Lahcen, R. B.: Liver cell adenomas and peliosis hepatis in mice associated with oxazepam. Res. Commun. Chem. Path. Pharmacol. *8*, 481 (1974)

Frederick, W. C., Howard, R. G., Spatola, S.: Spontaneous rupture of the liver in patients using contraceptive pills. Arch. Surg. (Chicago) *108*, 93 (1974)

Garson, O. M., Robson, M. K.: Studies in a patient with acute leukaemia after lysergide treatment. Brit. med. J. *2*, 800 (1969)

Girardi, A. J., Sweet, B. H., Slotnick, V. B., Hilleman, M. R.: Development of tumors in hamsters inoculated in the neonatal period with vacuolating virus, SV 40. Proc. Soc. exp. Biol. (N. Y.) *109*, 649 (1962)

Götz, H., Reichenberger, M.: Ergebnisse einer Fragebogenaktion bei 1670 Dermatologen der Bundesrepublik Deutschland über Nebenwirkungen bei der Griseofulvintherapie. Hautarzt *23*, 485 (1972)

Gordan, G. S., Greenberg, B. G.: Exogenous estrogens and endometrial cancer. Postgrad. Med. *59*, 66 (1976)

Greenblatt, M., Kommineni, V., Conrad, C., Wallcave, L., Lijinsky, W.: In vivo conversion of phenmetrazine into its N-nitroso derivative. Nature (New Biol.) *236*, 25 (1972)

Greenblatt, M., Mirvish, S. S.: Dose-response studies with concurrent administration of piperazine and sodium nitrite to strain A mice. J. nat. Cancer Inst. *50*, 119 (1973)

Gross, F.: Abschließende Stellungnahme zur Frage eines erhöhten Brustkrebsrisikos durch Rauwolfia-Präparate. Dtsch. Ärzteblatt *73*, 2370 (1976)

Hammond, E., C., Selikoff, I. J., Robitzek, E. H.: Isoniazid therapy in relation to later occurrence of cancer in adults and in infants. Brit. Med. J. *2*, 792 (1967)

Heinonen, O. P., Shapiro, S., Tuonimen, L.: Reserpine use in relation to breast cancer. Lancet *2*, 675 (1974)

Hermann, R. E., David, T. E.: Spontaneous rupture of the liver caused by hepatomas. Surgery *74*, 715 (1973)

Herrmann, E.: Die Bedeutung fortgesetzter Thiouracilmedikation für die Proliferation des Schilddrüsengewebes. Schweiz. med. Wschr. *81*, 1097 (1951)

Hershman, J. M., Givens, J. R., Cassidy, C. E., Astwood, E. B.: Long-term outcome of hyperthyroidism treated with antithyroid drugs. J. Clin. Endocr. *26*, 803 (1966)

Hillemanns, H. G.: Hormonale Kontrazeptiva und Krebsentstehung. In: Aktuelle Probleme auf dem Gebiet der Cancerologie. Lettré, H., Wagner, G. (eds.) Berlin-Heidelberg-New York: Springer, 1971, Vol. III pp. 158–169

Hoffmann, D., Rathchamp, C., Liu, Y. Y.: Chemical studies on tobacco smoke. On the isolation and identification of volatile and non-volatile N-nitrosamines and hydrazines in cigarette smoke. In: N-Nitroso Compounds in the Environment. Bogovski, P., Walker, E. A. (eds.) Lyon: International Agency for Research on Cancer, 1974, pp. 159–165

Hoover, R., Gray, L. A., Cole, P., MacMahon, B.: Menopausal estrogens and breast cancer. New Engl. J. Med. *295*, 401 (1976)

Horvarth, E., Kovacs, K., Ross, R. C.: Benign hepatoma in a young woman on contraceptive steroids. Lancet *1*, 357 (1974)

Howe, R.: Carcinogenicity of Alderlin (pronethalol) in mice. Nature (Lond.) *207*, 594 (1965)

Hoyumpa, A. M., Schiff, L., Helfman, E. L.: Budd-Chiari syndrome in women taking oral contraceptives. Amer. J. Med. *50*, 137 (1971)

Johnson, F. L., Feagler, J. R., Lerner, K. G., Majerus, P. W., Siegel, M., Hartmann, J. R., Thomas, E. D.: Association of androgenic-anabolic steroid therapy with development of hepatocellular carcinoma. Lancet *2*, 1273 (1972)

Juhasz, J., Balo, J., Kendrey, G.: Über die geschwulsterzeugende Wirkung des Isonicotinsäurehydrazid (INH). Z. Krebsforsch. *62*, 188 (1957)

Jung, H.-D.: Zur fraglichen Induzierung von Bronchial- und Hautkarzinom durch INH (Isoniazid) am Beispiel der Tuberculosis luposa. Z. Erkrank. Atmungsorg. *135*, 31 (1971)

Kelso, D. R.: Benign hepatomas and oral contraceptives. Lancet *1*, 315 (1974)

Knapp, W. A., Ruebner, B. H.: Hepatomas and oral contraceptives. Lancet *1*, 270 (1974)

Korpassy, B.: Tannins as hepatic carcinogens. Progr. Exp. Tumor Res. *2*, 245 (1961)

Laerum, O. D., Iversen, O. H.: Reticuloses and epidermal tumors in hairless mice after topical skin application of cantharidin and asiaticoside. Cancer Res. *32*, 1463 (1972)

Laska, E. M., Siegel, C., Meisner, M., Fischer, S., Wanderling, J.: Matched-pairs study of reserpine use and breast cancer. Lancet *2*, 296 (1975)

Lijinsky, W.: Reaction of drugs with nitrous acid as a source of carcinogenic nitrosamines. Cancer Res. *34*, 255 (1974)

Lijinsky, W., Conrad, E., van de Bogart, R.: Carcinogenic nitrosamines formed by drug nitrite interation. Nature (Lond.) *239*, 165 (1972)

Lijinsky, W., Greenblatt, M.: Carcinogen dimethylnitrosamine produced in vivo from nitrite and aminopyrine. Nature (New Biol.) *236*, 177 (1972)

Lijinsky, W., Singer, G. M.: Formation of Nitrosamines from tertiary amines and nitrous acid. In: N-Nitroso Compounds in the Environment. Bogovski, P., Walker, E. A. (eds.) Lyon: International Agency for Research on Cancer, 1974, pp. 111– 114

Loube, S. D., Ruirk, R. A.: Breast cancer associated with administration of spironolactone. Lancet *1*, 1428 (1975)

Lundsgaard-Hansen, P.: Zur Frage der Bedeutung der Thiouracilderivate für die Entstehung maligner Tumoren, insbesondere von Schilddrüsentumoren. Oncologia *9*, 33 (1956)

Mack, T. M., Henderson, B. E., Gerkins, V. R., Arthur, M., Baptista, J., Pike, M. C.: Reserpine and breast cancer in a retirement community. New Engl. J. Med. *292*, 1366 (1975)

Mauer, S. M., Lee, C. S., Najarian, J. S., Brown, D. M.: Induction of malignant kidney tumors in rats with streptozotocin. Cancer Res. *24*, 158 (1974)

Mays, E. T., Christopherson, W. H., Barrows, G. A.: Focal nodular hyperplasia of the liver. Possible relationship to oral contraceptives. Amer. J. clin. Path. *61*, 735 (1974)

Meadows, A. T., Naiman, M. D., Valdes-Dapena, M.: Hepatoma associated with androgen therapy for aplastic anemia. J. Pediat. *84*, 109 (1974)

Meyer, P., Livolsi, V. A., Cornog, J. L.: Hepatoblastoma associated with an oral contraceptive. Lancet *1*, 1387 (1974)

Michalowsky, R., Kudejko, T.: Karzinomentstehung mit Vitamin D_2 und Isonikotinsäurehydrazid behandeltem Lupus vulgaris. Derm. Wschr. *151*, 25 (1965)

Minder, W. H.: Der Grahamsche Schilddrüsentumor (sklerosierendes Mikrocarcinom der Struma basedowiana) und seine Beziehungen zur thyreostatischen Therapie. Schweiz. med. Wschr. *82*, 393 (1952)

Mirvish, S. S.: Kinetics of N-nitrosation reactions in relation to tumorigenesis experiments with nitrite plus amines or ureas. In: N-Nitroso Compounds. Analysis and Formation. Lyon: International Agency for Research on Cancer, 1972, pp. 104– 108

Mori, K., Yasuno, A., Matsumoto, K.: Induction of pulmonary tumors in mice by isonicotinic acid hydrazide feeding. Gann *51*, 83 (1960)

Naeim, F., Copper, P. H., Semion, A. A.: Peliosis hepatis. Possible etiologic role of anabolic steroids. Arch. Path. (Chicago) *95*, 284 (1973)

Nyfors, A.: Lupus vulgaris, isoniacid and cancer. Scand. J. resp. Dis. *49*, 264 (1968)

O'Fallon, W. M., Labarthe, D. R., Kurland, L. T.: Rauwolfia derivatives and breast cancer. Lancet *2*, 292 (1975)

O'Sullivan, J. P., Wilding, R. P.: Liver hamartomas in patients on oral contraceptives. Brit. med. J. *3*, 7 (1974)

Payne, R. L., Crane, A. R., Price, J. G.: Thiouracil and earcinoma of the thyroid. Surgery *22*, 496 (1947)

Peacock, A., Peacock, P. R.: The results of prolonged administration of isoniacid to mice, rats, and hamsters. Brit. J. Cancer *20*, 307 (1966)

Pemberton, J., De, J., Black, B. M.: The association of carcinoma of the thyroid gland and exophthalmic goiter. Surg. Clin. N. Amer. *28*, 935 (1948)

Pompe, K.: Einfluß von Isonicotinhydrazid auf die Lupuskarzinomentstehung. Derm. Wschr. *133*, 105 (1956)

Price, J. M., Biava, C. G., Oser, B. L., Vogin, E. E., Steinfeld, J., Leey, H. L.: Bladder tumours in rats fed cyclohexylamine or high doses of a mixture of cyclamate and saccharine. Science *170*, 1131 (1970)

Price, J. M., Morris, J. E., Lalich, J. J.: Evaluation of the carcinogenic activity of 5-nitrofuran derivatives in rats. Fed. Proc. *25*, 419 (1966)

Purves, H. D., Griesbach, W. E., Kennedy, T. H.: Studies in experimental goitre: Malignant change in a transplantable rat tumor. Brit. J. Cancer *5*, 301 (1951)

Reveno, W. S., Rosenbaum, H.: Observation on the use of antithyroid drugs. Ann. Intern. Med. *60*, 982 (1964)

Rudas, B.: Streptozotocin. Arzneimittel-Forsch. *22*, 830 (1972)

Rustia, M., Shubik, P.: Induction of lung tumors and malignant lymphomas in mice by metronidazole. J. nat. Cancer Inst. *48*, 721 (1972)

Schmähl, D.: Fehlen einer cancerogenen Wirkung von Artemesia Absinthium nach Fütterung an Ratten. Z. Krebsforsch. *61*, 227 (1956)

Schmähl, D.: Fehlen einer cancerogenen Wirkung von Cyclamat, Cyclohexylamin und Saccharin bei Ratten. Arzneimittel-Forsch. *23*, 1466 (1973)

Schmähl, D.: Carcinogenic hazards from drugs. In: Advances in Tumour Prevention, Detection and Characterization. Maltoni, C. (ed.) Amsterdam-New York: Excerpta Medica, 1974, Vol. II, pp. 34–40

Schmähl, D.: Experimental investigations with anticancer drugs for carcinogenicity with special reference to immunodepression. Recent Results Cancer Res. *52*, 18 (1975)

Schmähl, D.: Kanzerogene Substanzen. Dtsch. Ärzteblatt *74*, 1–4 (1977)

Schmähl, D., Habs, M.: Life-span investigations for carcinogenicity of some immune-stimulating, immunodepressive, and neurotropic substances in Sprague-Dawley-rats. Z. Krebsforsch. *86*, 77–84 (1976)

Schmähl, D., Krüger, F. W.: Fehlen einer synkarzinogenen Wirkung von Cyclamat bei der Blasenkrebserzeugung mit Butyl-butanol-nitrosamin bei Ratten. Arzneimittel-Forsch. *22*, 999 (1972)

Schmähl, D., Steinhoff, D.: Kanzerogene Wirkung von 1-(p-Carboxyäthylphenyl)-3,3-dimethyltriazen nach s. c. und i. v. Gabe bei Ratten. Arzneimittel-Forsch. *23*, 800 (1973)

Sellers, E. A., Hill, J. M., Lee, R. B.: Effect of iodide and thyroid on the production of tumours of the thyroid and pituitary by propylthiouracil. Endocrinology *52*, 188 (1953)

Speyer, E., Walther, L.: Berichte *63 B*, 852 (1930)

Staffurth, J. S.: Thyroid cancer after 131 I therapy for thyrotoxicosis. Brit. J. Radiol. *39*, 471 (1966)

Stenwig, A. E., Solgaard, T.: Ruptured benign hepatoma associated with an oral contraceptive. Virchows Arch. A. *367*, 337 (1975)

Tauntas, C., Paraskevas, G., Deligeorgi, H.: Benign hepatomas and oral contraceptives. Lancet *1*, 1351 (1974)

Taylor, H. W., Lijinsky, W.: Tumor induction in rats by feeding aminopyrine or oxytetracycline with nitrite. Int. J. Cancer *16*, 211 (1975)

109

Thomas, C., Rogg, H., Bücheler, J.: Die krebserzeugende Wirkung des N-Nitroso-methyl-harnstoffes nach Verabreichung hormonaler Kontrazeptiva. Beitr. Path. *146*, 332 (1972)

Torentas, C., Paraskevas, G., Deligeorgi, H.: Benign hepatoma and oral contraceptives. Lancet *1*, 1351 (1974)

Toth, B., Boreisha, J.: Tumorigenesis with isonicotinic acid and urethan in the Syrian golden hamsters. Europ. J. Cancer *5*, 165 (1969)

Urman, H. K., Bulay, O., Clayson, D. B., Shubik, P.: Carcinogenic effects of niridazole. Cancer Letters *1*, 69 (1975)

Walser, A., Fryer, R. I., Sternbach, L. H., Archer, M. D.: Quinazolines and 1,4-benzodiazepines. Some transformations of chlordiazepoxide. J. Heterocyclic Chem. *11*, 619 (1974)

Watanabe, F., Sugimoto, S.: Seven cases of transplantable sarcomas of rats appearing in the areas of repeated sc. injections of urotropine. Gann *46*, 365 (1955)

Zick, J., Armstrong, B.: Breast cancer and spironolactone. Lancet *2*, 368 (1975)

Ziel, H. K., Finkle, W. D.: Increased risk of endometrial carcinoma among users of conjugated estrogens. New Engl. J. Med. *293*, 1167 (1975)

Table 25. Drugs considered to be potential carcinogens in man on the basis of experimental results or case reports

Certain/probable	Possible	Improbable/not assessable
Alkylating agents (mustard gas derivatives, e. g., cyclophosphamide; ethyleneimines, e. g., thiotepa, triazenes, nitrosoureas)	Adriamycin	Cantharidin
	Antimetabolites	Chinoline derivatives
	Chloramphenicol	Cyclamate
	Griseofulvin	Herbae Artemesiae Absinthii
Arsenic	Halogenated paraffins	Hexamethylenetetramine
Diethylstilbestrol (transplacentally)	Hydantoin derivatives	Iron dextran
Procarbazine	Nitrofuran derivatives	Isonicotinic acid hydrazide
	Phenacetin	Lactams
	Phenylbutazone	Lysergides
	Streptozotocins	Metronidazole
	Tar salves	Niridazole
	Thiouracils	Polyvinylpyrrolidone and similar plasma expanders
	Urethan	Potassium perchloride
		Pronethalol
		Rauwolfia derivatives
		Saccharin
		Safrols
		Spironolactones
		Tannin (orally)

Table 26. Interactions of drugs with nitrite in chemical systems and in vivo (according to Lijinsky)

Drug	Chemical system	Biological system	Reference
Aminopyrine	+	+ *	Lijinsky et al., 1972; Lijinsky, 1974 Lijinsky and Greenblatt, 1972; Taylor and Lijinsky, 1975
Analgine	+		Eisenbrand, unpublished data
Chlordiazepoxide	+	+ ?	Walser, et al., 1974 Lijinsky and Taylor, unpublished data
Chlorpheniramine	+		Lijinsky, 1974
Chlorpromazine	+	+ −	Lijinsky, 1974 Lijinsky and Taylor, unpublished data
Cyclizine	+	+ ?	Lijinsky, 1974 Lijinsky and Taylor, unpublished data
Dextropropoxyphene	+		Lijinsky, 1974
Disulfiram	+		Lijinsky et al., 1972, Lijinsky, 1974
Hexamethylenetetramine	+	+ −	Bachmann and Deno, 1951 Lijinsky and Taylor, unpublished data
Lucanthone	+	+ −	Lijinsky, 1974 Lijinsky and Taylor, unpublished data
Methadone	+		Lijinsky, 1974
Methapyriline	+	+ ?	Lijinsky, 1974 Lijinsky and Taylor, unpublished data
Morphine	+		Speyer and Walther, 1930
Nicotine	+		Hoffmann et al., 1974
Nikethamide	+		Lijinsky et al., 1972
Oxytetracycline	+	+ ?	Lijinsky et al., 1972; Lijinsky, 1974 Taylor and Lijinsky, 1975

* = positive results in vivo (i. e., induction of tumors)
− = negative results in vivo
? = questionable results in vivo

Table 26 (continued)

Drug	Chemical system	Biological system	Reference
Phenacetin	+		Eisenbrand and Preussmann, 1975
Phenmetrazine	+	+ −	Greenblatt et al., 1972 Greenblatt et al., 1972
Piperazine	+	+ *	Mirvish, 1972 Greenblatt and Mirvish, 1973
Piperine	+		Lijinsky et al., 1972
Quinacrine	+		Lijinsky, 1974
Tolazamide	+	+ −	Lijinsky et al., 1972 Lijinsky, 1974 Lijinsky and Taylor, unpublished data

* = positive results in vivo (i. e., induction of tumors)
− = negative results in vivo
? = questionable results in vivo

112

Table 27. Reaction of some drugs with nitrite under "stomach-like" conditions (pH 1–2, 37° C) and formation of nitrosamines

Drug	Nitrosamine formed	Carcinogenicity of the nitrosamines	Yield of reaction
Aminopyrine (Pyramidon)	Dimethylnitrosamine	+ + +	High
Oxytetracycline	Dimethylnitrosamine	+ + +	Moderate
Chlorpromazine	Dimethylnitrosamine	+ + +	Small
Dextropropoxyphene	Dimethylnitrosamine	+ + +	Small
Methadone	Dimethylnitrosamine	+ + +	Small
Methapyriline	Dimethylnitrosamine	+ + +	Small
Lucanthone	Diethylnitrosamine	+ + +	Moderate
Quinacrine	Diethylnitrosamine	+ + +	Moderate
Disulfiram	Diethylnitrosamine	+ + +	Small
Nikethamide	Diethylnitrosamine	+ + +	Small
Tolazamide	N-nitrosohexamethyleneimine	+ +	Small
Cyclizine	Nitrosopiperazine + Dinitrosopiperazine	+ +	Small
Ephedrine	N-nitrosoephedrine	+ +	Moderate
Phenmetrazine	N-nitrosophenmetrazine	–	Moderate
Methylphenidate	N-nitrosomethylphenidate	–	Moderate
Phenacetin	N-nitroso-2-nitro-4-aethoxy-acetanilide	?	Moderate

Discussion

The examination and evaluation of drugs with regard to possible carcinogenic side-effects is a relatively new field of toxicology. Its importance has been increasing in recent years and possibly will increase more after the appearance of this book. Since the leading principle of each therapeutic measure should be *nil nocere,* particular attention should be paid to the long-term effects of drugs.

Adverse effects of a drug that occur shortly after its use are readily detected by common toxicologic methods. There are, however, side effects that occur only after a long delay. To these belong carcinogenic effects. We therefore would like to describe briefly the design and evaluation of long-term studies for carcinogenesis.

The first step is the choice of the drugs for carcinogenesis experiments. It is impossible and unrealistic to examine all drugs for carcinogenic effects. However, carcinogenesis experiments should be carried out when such effects are possible due to the chemical structure and reactivity of a substance (i.e., alkylating properties), or when a drug is recommended for prolonged use in children or pregnant women.

The studies should be carried out in at least two suitable animal species, which should be treated for their entire life. The mode of application should be orientated to that used in humans. Two dosages, one of which should be 50–100 times as large as the clinical dose, should be used. There are, however, a few exceptions to this rule. Alkylating agents, i.e., cyclophosphamide, act predominantly on the proliferating tissues. The common clinical single dose of cyclophosphamide is 2–3 mg/kg body weight. Acute fatal toxicity develops by administering a tenfold dose to the animals. We would thus not be able to conduct long-term toxicity studies with this substance. This example shows that the experimental doses of a drug have to be orientated to both the pharmacologic effects of the drug and to the clinical conditions. In most cases the study of the transplacental carcinogenic effects of a drug is unavoidable (for methods see Ivankovic, 1975).

The following examples, some of which were already dealt with in the text, may explain our ideas for the evaluation of drugs.

3-Phenyl-5-β-diethylamino-ethyl-1,2,4-oxadiazole, a derivative of oxadiazole, is an example of a substance that, after its toxicologic long-term examination, was not introduced into the market. It was initially recommended to be used as a cough remedy, but showed carcinogenic effects on the bladders of rats and dogs (Barron, 1963).

INH induced lymphadenomas, lymphosarcomas, and leukoses in mice after long-term application of small doses, whereas it was not effective in rats and hamsters. We have thus far no indications of carcinogenicity of INH in man; however, the period of observation is probably still too short to draw final conclusions. On the other hand, INH has such beneficial tuberculostatic effects that we can hardly renounce its use. This example may show that the therapeutic effect of a drug has to be balanced against its potential risk in each case. The synthesis of a substance with equal tuberculostatic effects but lacking carcinogenicity would be the most desirable solution in this case. Such a hope was recently expressed by Weisburger et al. (1975). However, we would like to emphasize that the separation of the carcinogenic component of a substance from its therapeutic component is extremely difficult.

The use of arsenic as a chemotherapeutic agent is another problem altogether. Carcinogenic effects of this substance have been known for more than 100 years; nevertheless, it is still used by some dermatologists and internal specialists. As already mentioned, there is no longer a need for medical use of arsenic; prescription of arsenic is therefore considered a malpractice.

In this book it has been shown that alkylating cytostatic agents used in cancer chemotherapy and immunodepression may have carcinogenic effects in animals and man. They should therefore be used only in extreme circumstances which of course include chemotherapy of malignant tumors. Postoperative tumor treatment ("adjuvant chemotherapy") with these compounds, however, should be done only in exceptional cases and should be avoided particularly when the operation provides a good chance of recovery, i.e. in earliest stages of mammary or cervical cancer. This view is taken by American colleagues also (Penn, 1974; Harris, 1976). We also warn against using cytostatic agents for treating benign hyperproliferations and in immunosuppression, where less hazardous alternatives are available.

The example of phenacetin clearly shows the difficulties in finding a reasonable compromise between experimental findings and clinical observations. This substance, which is often used in analgesic mixtures, led to carcinomas of the renal pelvis in analgesic abusers. In animal experiments, which were carried out because of its chemical similarity to the sweetener dulcin, phenacetin did not show carcinogenicity although very large doses were given. In this particular case man may be more sensitive to the carcinogenic action of this substance than

115

experimental animals. The problem of phenacetin-related carcinoma of the renal pelvis should be kept in mind in order that the drug might be withdrawn from the market, should further observations be reported. The dose in this case may be the most decisive factor, since kidney carcinomas were observed only in real phenacetin abusers.

The last example that one can cite is cyclamate, which, on the basis of dubious experimental findings, was banned in the USA and in several European countries. Numerous experimental studies carried out in various countries have since contradicted the above findings. The example of cyclamate shows that, unless the carcinogenicity of a substance is proved by reproducible results, such a chemical, which is extremely valuable otherwise, should not be banned.

The present clinical and experimental data clearly show that a few drugs may have carcinogenic effects in man. The pharmacotherapist therefore should be familiar with the fundamental publications on this subject (Truhaut, 1967; Clayson, 1972; Schmähl, 1972, 1974) in order to balance the benefit of a drug against its potential risk in individual cases.

Three medical branches contribute predominantly to the detection of carcinogenic side-effects of medical intervention. The tasks of experimental toxicology have been pointed out before. The clinicians are requested to use possibly carcinogenic drugs with respective caution and to study extensively the case history of a patient in order to associate a tumor disease with the use of a certain therapy. Pathology is mentioned because not seldom associations between inadequate therapy and the formation of a tumor were detected by post-mortem examination.

We appeal to all physicians to pay particular attention to iatrogenic carcinogenesis in order to protect the patient from possible injury.

References

Barron, C. N.: Observations on the chronic toxicity of 3-phenyl-5-β-diethylamino-ethyl-1,2,4-oxadiazole in the rat and dog. Exp. Molec. Path. 2, 1 (1963)

Clayson, D. B.: Carcinogenic hazards due to drugs. In: Drug-Induced Diseases. Meyler, L., Peck, H. M. (eds.) Amsterdam: Excerpta Medica 1972, Vol. IV, pp. 91–109

Harris, C. C.: The carcinogenicity of anticancer drugs: A hazard in man. Cancer (Philad.) 37, 1014 (1976)

Ivankovic, S.: Praenatale Carcinogenese. In: Handbuch der allgemeinen Pathologie. Geschwülste/Tumors III. Altmann, H.-W., Büchner, F., Cottier, H., Grundmann, E., Holle, G., Letterer, E., Masshoff, W., Meesen, H., Roulet, F., Seifert, G., Siebert, G. (eds.) Berlin-Heidelberg-New York: Springer 1975, Vol. 6, pp. 941–1002

Penn, I.: Chemical immunosuppression and human cancer. Cancer (Philad.) *34,* 1474 (1974)

Schmähl, D.: Toxikologische Probleme der iatrogenen Carcinogenese. Verh. dtsch. Ges. Path. *56,* 133 (1972)

Schmähl, D.: Carcinogenic hazards from drugs. In: Cancer Detection and Prevention. Maltoni, C. (ed.) Amsterdam: Excerpta Medica 1974, pp. 34–40

Truhaut, R. (ed.): Potential carcinogenic hazards from drugs. UICC Monograph Series, Vol. VII, Berlin-Heidelberg-New York: Springer 1967

Weisburger, J. H., Griswold, D. P., Prejeau, J. D., Casey, A. E., Wood, H. B., Weisburger, E. K.: The carcinogenic properties of some of the principal drugs used in clinical cancer chemotherapy. Rec. Results Cancer Res. *52,* 1 (1975)

See also the just published IARC Monograph Series on the Evaluation of Carcinogenic Risk of Chemicals to Man. Vol. XIII. Some Miscellaneous Pharmaceutical Substances. Lyon: International Agency for Research on Cancer, 1977.

Subject Index

Recent Results in Cancer Research

Sponsored by the Swiss League against Cancer. Editor in Chief: P. Rentchnick, Genève